A Clinical Companion to Biochemical Studies

A Clinical Companion to Biochemical Studies

Victor Schwarz
UNIVERSITY OF MANCHESTER

W. H. FREEMAN AND COMPANY
Reading and San Francisco

Library of Congress Cataloguing in Publication Data

Schwarz, Victor, 1920–
 A clinical companion to biochemical studies.

 1. Physiology, Pathological—Cases, clinical
reports, statistics. 2. Biological chemistry—Cases,
clinical reports, statistics. I. Title.
[DNLM: 1. Biochemistry. QU4 S411c]
RB113.S39   616.07   77–17132

ISBN 0–7167–0078–6
ISBN 0–7167–0077–8 pbk.

Phototypeset in 11 pt V.I.P. Melior by Western
Printing Services Ltd, Bristol
Printed in the United States of America

To Mary, Isabel and Elena

Foreword

Today's medical students have a hard life. New methods of diagnosis and treatment appear all the time and the medical sciences grow at an alarming rate. I doubt if there is a medical curriculum in any university which is not overcrowded and yet each professor will assure the students that his particular discipline is essential.

Over the past thirty years Biochemistry has grown faster than any of the other basic medical sciences. One look at the chart of metabolic pathways supplied by a friendly manufacturer of biochemicals will deter all but the strongest. "Do I need to know the whole of that?" asks the medical student, who wants to be curing diseases rather than contemplating molecules. Of course he does not need to know it all, but he does need to feel at home with the major pathways. Biochemistry will contribute to the understanding and treatment of a surprising number of illnesses and the number grows each year.

There are several excellent text books of biochemistry for the science student, though I am less happy with those written for medical students. There are also many books on clinical biochemistry, but these are usually written for laboratory workers rather than students.

Dr. Schwarz's book breaks new ground, since it presents clinical biochemistry to the medical student without pretending to be a full text book of biochemistry. It shows medical students how important the subject is for clinical medicine today and does this in a clear and interesting way. In my view every medical student should have at least two biochemistry texts; this one, and a comprehensive yet readable text book on the subject, not necessarily one designed for medical students.

There is at present a shortage of medical graduates with an interest in clinical biochemistry. At the same time, many of the research proposals put to medical research foundations depend on a deep knowledge of biochemistry. I believe that this book will help to solve this problem. If only a few of those who read it decide to pursue biochemistry seriously it will have been extremely useful in the progress of medicine.

January 1978   J. N. Hawthorne
Professor of Biochemistry and Vice-Dean,
Queen's Medical Centre, Nottingham

Contents

x *Contents*

Preface

In most medical courses the basic pre-clinical sciences are taught in the first 4–6 terms, during which diseases may be referred to briefly and patients brought into the lecture theatre for demonstration. This latter practice defers to students' eagerness to come to grips with 'real' medicine yet fails to achieve its primary objective—to provide the stimulation and motivation to guide the student through the maze of pre-clinical studies. Such support is invaluable in helping to dispel doubts in the student's mind regarding the relevance of the biochemistry course, doubts which inhibit his learning and sap his determination to embrace material which appears academic or at best peripheral.

In order to counteract this negative attitude, we have, for the past two years, given a series of 'clinical biochemical seminars' during the first 4 terms of our pre-clinical course. In each seminar one or two case histories were presented which contained the most significant, representative and biochemically relevant clinical material pertaining to the particular disease. The presentation of each case was followed by an examination of the underlying biochemical mechanisms and, wherever possible, explanations were offered for the clinical observations. The seminars have proved popular with students and staff alike and have stimulated much useful discussion. I hope that this book may help a wider circle of students to gain insight into the close relationship between biochemistry and clinical medicine, which so often eludes them in the early years of their medical studies.

Many of the cases presented here are substantially in the form in which we have discussed them in our seminars. Others have been added to cover material often taught later in the second year. All have been chosen to provide the widest possible range of biochemical topics with a minimum of overlap. Carbohydrate, amino acid, triglyceride, haemoglobin, porphyrin and nucleic acid metabolism, protein structure, enzyme action and co-enzymes, ion transport and the biosynthesis and biochemical effects of hormones are represented.

It has not been my aim to be comprehensive, either in the description of the disease or even in the discussion of the biochemistry, as this could so easily defeat the whole purpose of the venture; nor is there any relationship between the choice of diseases and their incidence or 'importance'.

Sufficient detail has been given to enable the student to understand the underlying biochemical basis of the disease as well as his, or indeed available, knowledge permits, without reference to other works. But, as its title suggests, this book is intended to be read in conjunction with biochemistry texts. The order of individual cases is roughly in accordance with the clinical and biochemical complexity of the material.

Inherited enzyme defects are particularly well represented. These 'inborn errors of metabolism' have been described as nature's perfect experiments in biochemistry, which cannot be simulated in the laboratory. The deletion of one particular enzyme shows us not only the importance of the particular step in a metabolic sequence but also the often unsuspected effects of the accumulating intermediates on other enzymes, and hence on the functioning of the cell and of the whole body. There is no clearer or more convincing way of demonstrating the biochemical basis of many pathological lesions, from abnormal shapes of red blood cells, derangement of liver function, cataracts and aberrations of sex organs to mental defect.

My sources have been textbooks, monographs and original papers, from which I have synthesized case histories to make them, as far as possible, typical as well as subservient to my purposes. A few references are given to enable those who desire further insight to gain access to the literature.

I would like to record my indebtedness to my colleagues in the Department of Medical Biochemistry and in other pre-clinical and clinical departments of the University, who have made many valuable suggestions incorporated in the text.

I am grateful to the following for providing the photographs: Drs D. I. K. Evans, G. M. Komrower and I. B. Sardharwalla, Royal Manchester Children's Hospital; Prof. J. H. Kellgren, Drs J. E. MacIver, S. Oleesky and B. R. Tulloch, Manchester Royal Infirmary and Prof. Z. Lojda, First Pathological Institute, University of Prague; also to the University Department of Medical Illustration, Manchester Royal Infirmary, for the preparation of photographs and for permission to reproduce them.

V.S.

Manchester, April 1977

Introduction

Learning without thinking is useless.
Thinking without learning is dangerous.
Confucius

The brief case histories here presented are, as far as possible, typical of the particular diseases; they have been synthesized from cases described in the medical literature in greater detail. Some signs and symptoms have been omitted for the sake of simplicity, since this book is intended for students who have not yet studied clinical medicine; but nothing that is included is irrelevant to my primary aim, which is to generate interest in biochemistry by drawing attention to its fundamental role in the disease process, in diagnosis and treatment.

The reader should give careful consideration to every detail of the case history and should ponder, rather than commit to memory, the biochemical and other data in tables and figures, which have been selected for their pertinence and their instructive value. As far as existing knowledge, and space, permit, the phenomena described have been explained in the discussion.

Understanding the molecular mechanisms underlying disease not only lends depth to clinical studies but offers a wealth of intellectual satisfaction and stimulation to the student of biochemistry.

Some questions at the end of each chapter are offered as a challenge; the answers may be contained in the text or they may be gained from perusal of biochemistry books. Many more questions will present themselves to the thoughtful reader and will, I hope, stimulate his curiosity further to explore the subject. A few selected references are given for those whose interest and available time allow them to study the subjects discussed in greater biochemical and clinical detail.

Temporary Lactase Deficiency

Case History Brenda J. was born after a normal pregnancy and delivery. She was breast-fed and was in good health until she was weaned at 2 months. A cow's milk preparation was then introduced. Three days later Brenda developed diarrhoea, she became irritable, vomited frequently and had a temperature of 40.5 °C. She was admitted to hospital with a diagnosis of gastro-enteritis. She was treated for dehydration with intravenous fluids for 24 h, followed by glucose solution by mouth for another 2 days. Vomiting and diarrhoea stopped and Brenda was much more settled. She gained weight; the urine was normal and contained no reducing substance. Her gastro-enteritis had probably been caused by inadequate attention to hygiene in the preparation of the milk.

On her 4th day in hospital Brenda was given a brand of infant milk which she took avidly; but after 24 h her stools became loose and frequent. The following day her urine contained a reducing sugar which was not glucose, as indicated by a negative reaction on the glucose oxidase test strip. The stools were watery, frothy and acidic, with much gas being produced (Table 1); they also contained a reducing substance.

In the light of these findings and pending further investigations it was felt that Brenda's trouble might be a temporary lactase deficiency and she was put on a lactose-free diet. Her condition improved dramatically, the diarrhoea ceased, the reducing substance disappeared from her urine and stools, and she gained weight.

Meanwhile the reducing substance in the urine was identified as lactose. In an absorption test the patient was given an oral dose of the disaccharide and her plasma glucose and galactose was determined at intervals (Fig. 1); the plasma sugars were also chromatographed on paper (Fig. 2a). An intestinal biopsy specimen failed to stain for lactase. These four tests together indicated a failure to hydrolyse lactose due to the absence of lactase from the brush border of the intestinal mucosa.

After 6 weeks on a milk-free diet the lactose absorption test was repeated and found to be normal and a sugar chromatogram showed the presence in the plasma of galactose and the absence of lactose (Fig. 2b). A second

Table 1. Nature of stools. (Patient on milk diet)

Frequent
Watery
Frothy
pH 5
Contain a reducing substance and volatile acids

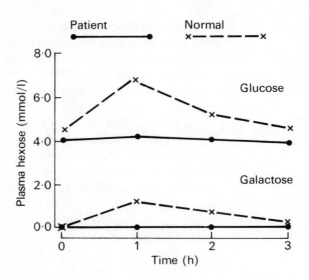

Fig. 1. Lactose absorption test. A standard amount of lactose is given by mouth and the plasma glucose and galactose levels are determined at intervals

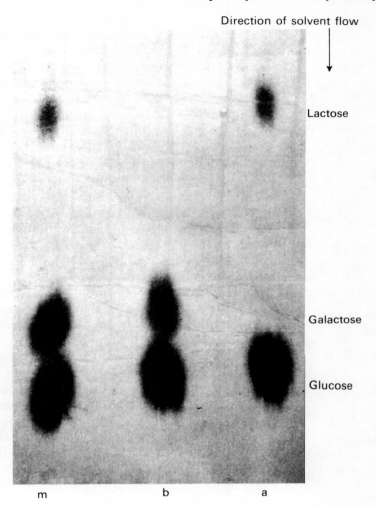

Direction of solvent flow

Lactose

Galactose

Glucose

Fig. 2. Chromatogram of sugars extracted from plasma after an oral lactose load. (m) Lactose, galactose and glucose markers, 10 μg each. (a) Patient's plasma, first absorption test; (b) patient's plasma, 6 weeks later

m   b   a

mucosal biopsy gave a normal reaction to lactase staining (Plate 1), confirming that the enzyme had returned. Thereafter milk did not cause any further gastro-intestinal disturbances.

Discussion

This child obviously did not suffer from an hereditary lactase deficiency, since she had thrived on breast milk for the first 2 months of life. She had developed gastro-enteritis when bottle feeding was begun and, as a result of the infection and the consequent malnutrition, the brush border lactase had disappeared. When she eventually took her milk feeds again, lactose failed to be hydrolysed and

Fig. 3. Action of
intestinal lactase

$$\text{Lactose} \xrightarrow[\text{(brush border)}]{\textit{lactase}} \text{glucose} + \text{galactose}$$

thus remained largely unabsorbed, only a small proportion passing into the circulation to be excreted in the
urine. (There is no provision for the hydrolysis and
metabolism of circulating lactose and hence the disaccharide, once absorbed, is excreted in the kidney.)

The unabsorbed lactose attracts water into the lumen of
the gastro-intestinal tract, so accounting for the watery
stools, and the presence of a fermentable sugar in the large
bowel results in proliferation of the micro-flora and its
metabolism of lactose to lactic and other acids and to CO_2.

In the presence of lactase, lactose is hydrolysed to
glucose and galactose (Fig. 3), which are absorbed and can
be detected in the plasma in the course of the absorption
test. Lactase is one of several disaccharidases in the brush
border and it appears to be specially sensitive to damage
by infections and other processes, not only in infants but
also in adults.

Many non-Europeans lose their intestinal lactase per-
manently after infancy and hence do not tolerate milk. An
enquiry in an American city revealed that 40 per cent of
lactase-deficient subjects were unaware of their deficiency. A Sudanese doctor described his gastro-intestinal
symptoms of many years' standing, which were eventually diagnosed as being due to his inability to metabolize
lactose (Ahmed, 1975).

Further Reading

Ahmed, H. F. (1975) *Lancet i*, 319.
Gray, G. M. (1972) in *The Metabolic Basis of Inherited
Disease* (Stanbury, J. B., Wyngaarden, J. B. & Frederickson,
D. S., eds), McGraw-Hill, New York, pp. 1453–64.

Questions

1. On the chromatogram (Fig. 2) 10 μg of each sugar was
applied as a marker. Why is the spot produced by lactose
smaller than those due to glucose and galactose?

2. How can one account for the low pH and the frothy
nature of the stools?

3. Why are the symptoms in adults relatively mild compared to those of Brenda J.?

Fructose Intolerance

Case History

Andrew K. was born after a normal pregnancy and was breast-fed until he was 7 months old. His mother then put him on a cow's milk preparation to which she added sucrose. Andrew responded to the changed diet by vomiting and refusing his feeds. By trial and error his mother eventually established that it was sweet foods that made him vomit and weaned him on to a mixed diet without any sugar. Thereafter he continued to develop normally. When he was 9 years old his teeth were in excellent condition, with no evidence of caries.

At that time a brother, Philip, was born, who was weaned of the breast after only 5 days, to be given an infant milk to which the mother added cane sugar. Philip vomited repeatedly, began to lose weight and became lethargic. On admission to hospital he was sweating, he had convulsions and his reflexes were absent, his liver was enlarged and he was jaundiced. His blood sugar, determined by a reducing-sugar method, was in the normal range, but by glucose oxidase it was found to be very low. The plasma inorganic phosphate concentration was also low (Table 2). The urine contained a reducing sugar, which was identified as fructose by paper chromatography.

Rapid clinical improvement followed the withdrawal of sugar from Philip's diet. Fructose disappeared from the blood and urine, the blood glucose returned to normal, the

Table 2. Blood sugar and inorganic phosphate

Metabolite	Amount in plasma	
	Patient	Normal range
	mmol/l	
Total reducing sugar	5.3	4.0–6.3
True glucose (by glucose oxidase)	0.4	3.3–5.6
Inorganic phosphate	0.4	1.5–1.9

jaundice faded and the liver regressed to its normal size. His weight increased rapidly and his further development was uneventful. A fructose tolerance test, carried out 2 weeks after eliminating fructose from the diet, showed a marked fall of blood glucose and serum inorganic phosphate concomitant with the rise in blood fructose (Fig. 4), which confirmed the diagnosis of fructose intolerance.

Fig. 4. Fructose tolerance test

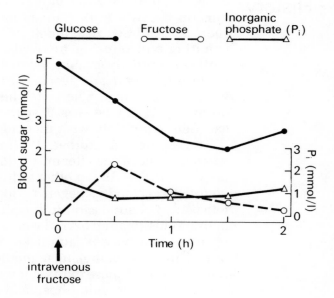

Discussion The profound hypoglycaemia, evident in the untreated patient and responsible for the sweating, convulsions and absent reflexes, is not due to release of insulin triggered by the elevated blood fructose concentration; the ketohexose does not stimulate the β-cells of the pancreas. The blood insulin level remained constant during the fructose tolerance test while the glucose fell from 4.8 to 2 mmol/l. Moreover, the fructose-induced hypoglycaemia does not respond to glucagon, which normally accelerates glycogenolysis and thereby counteracts the effect of insulin.

The defective enzyme in fructose intolerance is fructose-1-phosphate aldolase, whose action is very similar to that of the aldolase on the Embden-Meyerhof pathway but yields unphosphorylated glyceraldehyde. In the absence of this enzyme fructose 1-phosphate accumulates in the liver cells where it inhibits the aldolase-catalysed condensation of triose phosphate to fructose 1,6-diphosphate. This inhibition precludes formation of glucose from glycerol, lactic acid or amino acids by the normal gluconeogenetic pathway (Fig. 5). A further interference with carbohydrate metabolism occurs at the level of

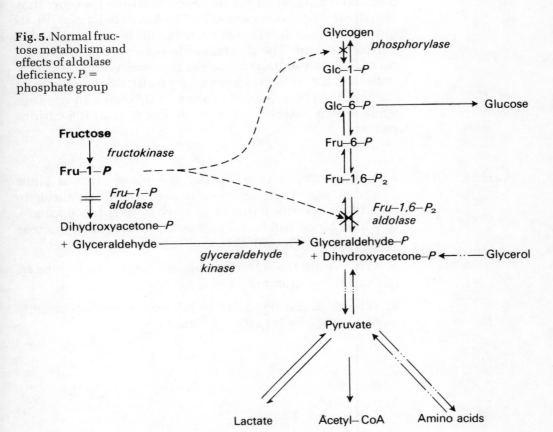

Fig. 5. Normal fructose metabolism and effects of aldolase deficiency. P = phosphate group

phosphorylase, possibly as a result of an inadequate supply of inorganic phosphate for the production of glucose 1-phosphate. This, together with inhibition of gluconeogenesis, accounts for the hypoglycaemia.

The hypophosphataemia is almost certainly related to the large amount of phosphate sequestered in the form of fructose 1-phosphate in the liver.

Liver damage is indicated by jaundice and other signs; it is due to the accumulation of fructose 1-phosphate in the cells and, probably, to a disturbance of their energy metabolism.

The defect in the fructose-1-phosphate aldolase is inherited and two siblings are affected in the K. family. No diagnosis was made on Andrew, since he was breast-fed for a much longer period and his mother withheld sugar when she found that it did not suit him. In the absence of fructose the genetic defect was not detected but Andrew's teeth bore witness to his mother's intuitive therapy: they remained free from caries. In Western society, with its inordinate consumption of sucrose, this is most unusual in a nine-year-old. The disaccharide is converted by mouth bacteria into dextran, a gel-like material which, together with salivary breakdown products, forms dental plaque on the tooth surface. Bacteria thrive in this gel and produce acids which dissolve the enamel and so start the carious process.

Questions

1. Neither hypoglycaemia nor the other clinical signs described above are observed in essential fructosuria, an hereditary condition due to absence of liver fructokinase, in which blood fructose levels as high as those in fructose intolerance are found. Explain.

2. Why would the activity of phosphorylase be limited by the supply of inorganic phosphate?

3. Why does glucagon fail to relieve the hypoglycaemia provoked by the ingestion of fructose?

Galactosaemia

Case History

Angela K. was born to healthy parents after a normal delivery in hospital on 3 March 1972. She was breast-fed but on the second day she refused her feed and vomited several times. The following day she developed diarrhoea and she lost weight. She continued to be fretful, taking her feeds reluctantly and vomiting frequently, and on 8 March was found to be jaundiced. In the course of the next few days the jaundice deepened, with both conjugated and unconjugated bilirubin in the serum. The liver was enlarged. Angela was put on half-strength modified cow's milk on which she seemed to improve somewhat, although vomiting and diarrhoea continued. On 16 March cataracts were noticed in her lenses. The reducing sugar in the blood was raised, and the glucose (determined by glucose oxidase) was low (Table 3). The urine gave a positive test for reducing sugar, which on chromatography was shown to be galactose. It also contained amino acids and protein (Table 4). A diagnosis of galactosaemia

Table 3. Patient's blood sugar

Sugar	Patient	Normal range
	mmol/l	
Total reducing sugar	9.0	4–6.3
Glucose	2.9	3.3–5.6

Table 4. Summary of clinical and biochemical features

Refusal of feeds
Vomiting and diarrhoea
Liver enlargement and jaundice
Blood reducing sugar raised, glucose low
Cataracts
Urine contains galactose, amino acids and protein

was made and Angela was put on a low-lactose diet on 18 March. Within 2 days the vomiting and diarrhoea stopped and the galactose disappeared from the blood and urine. She took her feeds well and began to put on weight. The jaundice faded and by the end of March her liver had returned to its normal size. Thereafter she continued to make good progress, but her cataracts remained.

Discussion

Galactose is metabolized, mainly in the liver, according to the scheme shown in Fig. 6.

In hereditary galactosaemia of this type there is a deficiency of uridyl transferase in the liver and in other tissues, which can be conveniently demonstrated in lysed red cells (Table 5): little or no transferase activity is found

Fig. 6. Metabolism of galactose. Gal-1-P, galactose 1-phosphate; Glc-1-P, glucose 1-phosphate

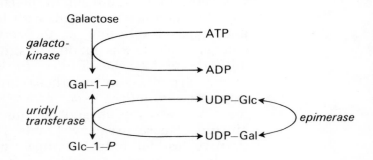

Table 5. Activity of uridyl transferase in haemolysates (UDP-Glc = uridine diphosphate glucose)

Source	Uridyl transferase activity ±S.D.
	μmol UDP-Glc h^{-1} g haemoglobin^{-1}
Patient	0
Normal children or adults	5.9 ± 1.0*
Parents of galactosaemic children	2.8 ± 0.7*

*Donnell, G. N., Bergren, W. R., Bretthauer, R. K. and Hansen, R.G. (1960) *Pediatrics 25,* 572.

in the patient's cells, while the parents' cells contain about half of the normal amount. The parents are heterozygous, a state characterized by the presence in every cell of one normal gene for transferase and one defective gene, each making their respective proteins, one active, the other inactive.

When uridyl transferase is absent, as in this patient, galactose 1-phosphate accumulates in all tissues and is believed to be responsible, directly or indirectly, for the malfunction of the gastro-intestinal tract, liver, kidney and brain, and ultimately for the death of the patient, if the condition is not diagnosed and treated in time. However, this accumulation of galactose 1-phosphate is not the only metabolic disturbance caused by the absence of trans-ferase: galactose, too, accumulates, first in the blood where it accounts for the elevated 'reducing sugar' and from where it spills over into the urine, and secondly in the tissues. The lens is especially sensitive to relatively high concentrations of galactose: the hexose interacts with NADPH, catalysed by aldose reductase to form the cor-responding sugar alcohol (Fig. 7). It is the accumulation of

Fig. 7. Reduction of galactose

$$\text{Galactose} + \text{NADPH} + \text{H}^+ \xrightarrow[\textit{aldose reductase}]{} \text{galactitol} + \text{NADP}^+ \text{(dulcitol)}$$

galactitol (dulcitol) and the consequent depletion of NADPH in the lens which leads to damage to the structure of the tissue and to opacification (cataract). Aldose reduc-tase has a high K_m, i.e. a low affinity for the sugar. This is an interesting example of the biological advantage of en-zymes with high K_m: a small amount of galactose is con-stantly present in the circulation of the normal milk-fed infant and hence would tend to deplete NADPH and so lead to cataract. Moreover the same enzyme also reduces glucose, with similar consequences to the tissue level of the coenzyme. The high K_m ensures that at physiological concentrations of the two sugars the enzyme is inactive and essential supplies of NADPH are safeguarded. In diabetes, however, glucose concentrations reach or exceed the K_m of aldose reductase and thus give rise to diabetic cataract.

Is it to be assumed, then, that the damage suffered by the

Fig. 8. Accumulation of metabolites in galactokinase and uridyl transferase deficiencies. The metabolites accumulating are indicated in bold type

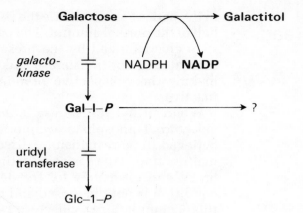

lens is due to depletion of NADPH by the aldose reductase reaction, whereas in all other tissues galactose 1-phosphate is the damaging agent? (Fig. 8). A comparison of the clinical symptoms of transferase deficiency and another type of galactosaemia—galactokinase deficiency—suggests that this is the case (Table 6).

Table 6. Effects of enzyme deficiency

Galactokinase	Uridyl transferase
Galactose in blood and urine	galactose in blood and urine
Cataracts	cataracts
	dysfunction of liver, kidney, spleen, intestine and brain
	death

The galactosaemic infant reared on a low-galactose diet develops normally, provided that no lasting damage has been suffered by the brain or the lenses. This illustrates the importance of the environment in either manifesting or concealing a genetic defect. The heterozygotes are clinically normal, despite the fact that they have only half the usual complement of transferase. But here again the environment is crucial, because the amount of galactose which can be tolerated is limited: a pint of milk is liable to produce mild gastric discomfort. In contrast, the homozygote galactosaemic patient has to beware even of the smallest quantities of dairy products.

Further Reading

Hansen, R. G. (1969) Hereditary galactosaemia, *J. Am. Med. Assoc. 208*, 2077–87.

Questions

1. Why did the galactose disappear from the blood and urine when the patient was given a low-lactose diet?

2. The lens capsule is not readily permeable to galactitol. What physical changes would you expect as a result of accumulation of the hexitol in the lens?

3. An infant showing the same symptoms as Angela K. is found, in laboratory tests, to have normal levels of galactokinase and uridyl transferase in the red cells. What conclusions would you draw?

Glycogen Storage Disease

Case History

David M. was born after an uneventful pregnancy. His mother noticed his large, protuberant abdomen soon after birth. In the course of the first few months he had frequent bouts of excessive sweating, pallor and generalized weakness, which were relieved by eating. When he was 12 months old he was examined in hospital and found to be mentally retarded. Blood glucose, after a 6-h fast, was low

Table 7. Fasting plasma levels of various metabolites

Metabolite	Level	
	Patient	Normal range
	mmol/l	
Glucose	1.6	3.3–5.6
Free fatty acids	1.6	0.3–0.9
Triglycerides	9.5	<1.4
Cholesterol	9.0	3.1–5.5
Uric acid	0.6	0.12–0.30

while the plasma lipids and uric acid were raised (Table 7). Liver biopsy showed extensive deposition of glycogen (Plates 2 and 3), which enabled a diagnosis of glycogen storage disease to be made.

Discussion The protuberant abdomen was due to enlargement of the liver, which accompanies excessive infiltration of glycogen. The connection between the massive glycogen deposition and the low fasting blood glucose (hypoglycaemia) was investigated in liver tissue, obtained at laparotomy. The three main enzymes concerned with the degradation of glycogen to glucose were determined (Fig. 9). Phosphorylase and debranching enzyme levels were within normal limits, whereas the glucose-

Fig. 9. Degradation of glycogen to glucose

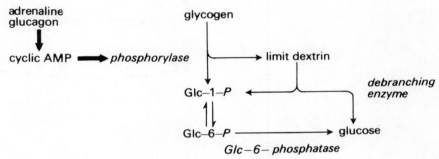

6-phosphatase activity was only about 10 per cent of the mean control value (Table 8). The formation of free glucose in the liver, and hence the maintenance of the blood glucose level, depends on this enzyme. Normally, when the blood glucose falls, phosphorylase is activated by glucagon or adrenaline to break down glycogen to glucose 1-phosphate which, after conversion to glucose 6-phosphate, yields free glucose by hydrolysis catalysed by glucose-6-phosphatase.

In this disease the patient's phosphorylase responds normally to glucagon and adrenaline, producing more glucose 1-phosphate and glucose 6-phosphate, which can,

Table 8. Glucose-6-phosphatase in liver. Measured as phosphate liberated from glucose 6-phosphate/g N in 30 min

Source	Glucose-6-phosphatase activity
	mg P_i/g N
Patient	22.4
Normal range	150–270

however, be hydrolysed only very slowly by the deficient enzyme. Instead, accumulation of this intermediate leads to increased formation of fructose 6-phosphate and thence, further down the glycolytic pathway, of pyruvate and lactate. In fact the patient's resting plasma levels of the

Fig. 10. Plasma glucose, lactate and pyruvate: effect of adrenaline

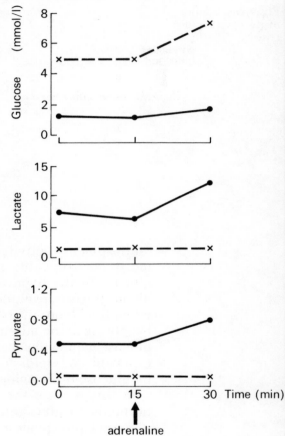

two acids were raised above those of control subjects and, when he was given adrenaline, his plasma glucose hardly changed but his lactate and pyruvate rose, in marked contrast to the control (Fig. 10). It is clear, therefore, that hypoglycaemia in this patient does call forth the remedial hormonal intervention, but the result is a greater output of lactate by the liver rather than of free glucose.

Continued production of adrenaline by the adrenal medulla is a potent stimulus to the adipose tissue to release free fatty acids and glycerol for oxidation and generation of energy, which is unobtainable from glucose. Thus the high free fatty acid level found in the patient's plasma probably reflects mobilization from the periphery. Other observations referable to adrenaline are the episodes of pallor and of excessive sweating.

The raised plasma levels of lipids and uric acid shown in Table 7 are further manifestations of disturbed metabolism. Since glycogenolysis cannot produce more than a limited amount of glucose, it floods the other pathways of glucose 6-phosphate metabolism. We have already seen how this leads to increased production of pyruvate and lactate, despite the allosteric mechanisms which control glycolysis. Some of the pyruvate will be metabolized to acetyl-CoA, not needed for oxidation and hence available for the synthesis of fat and cholesterol. The co-enzymes required for this synthesis, NADH and NADPH, are produced abundantly by the glycolytic and pentose phosphate pathways and hence all the precursors of triglyceride and cholesterol are readily available.

The pentose phosphate pathway, initiated by the oxidation of glucose 6-phosphate, yields (apart from NADPH) ribose 5-phosphate which stimulates, by a mass-action effect, the synthesis of phosphoribosylpyrophosphate and hence of purines. In this fashion overproduction of uric acid, the ultimate breakdown product of purines, could be explained (Fig. 11). The abnormally high plasma uric acid level may also be partly due to the raised plasma lactate, which interferes with the excretion of uric acid by the kidney. An extraordinary consequence of these metabolic disturbances, stemming from the deficiency of the distant glucose-6-phosphatase, is the development of the typical symptoms of gout due to the elevated uric acid concentration in the plasma.

How can the enzyme deficiency account for the accumulation of liver glycogen? Since glycogenolysis does not alleviate the hypoglycaemia, the adrenal cortex releases cortisol to promote gluconeogenesis—the formation of glucose from amino acids. In the face of the severe glucose-6-phosphatase deficiency little free glucose can be formed and the glucose 6-phosphate derived from amino acids is diverted to glycogen synthesis.

It is remarkable that the absence of a single enzyme

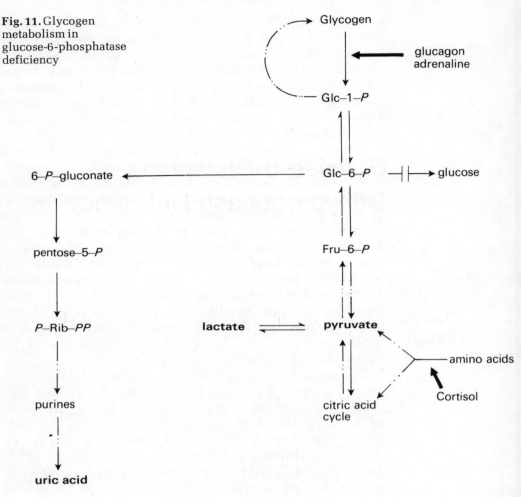

Fig. 11. Glycogen metabolism in glucose-6-phosphatase deficiency

leads, in this disease, to overproduction of glycogen, fat, cholesterol, purines and lactic acid, as well as to disturbances of brain, liver and kidney function.

Questions

1. What factors normally control the rate of glycolysis?

2. In which tissues does glucose-6-phosphatase occur?

Glucose-6-Phosphate Dehydrogenase Deficiency

Case History

R.P., a 19-year-old male native of Nigeria, was admitted to hospital with suspected anaemia and jaundice. He had left Africa recently and two weeks after his arrival in England he had developed a persistent headache, anorexia and nausea, and severe backache. He had bouts of shivering, a temperature of 40.6°C and he was delirious. Malaria was suspected and subsequently confirmed by identification of the parasite in a blood smear. The patient was prescribed primaquine by his doctor. Three days later he noticed that his urine was dark and the following day it was almost black. He complained of weakness and abdominal and back pain. His sclerae were yellow.

On examination in hospital R.P. was found to be weak and anorexic, with persistent vomiting. Haematological examination yielded the data presented in Table 9. The bilirubin was unconjugated. The red cells, on microscopic

Table 9.
Haematological data

	Patient	Normal range
Haemoglobin (g/dl)	9.2	14–18
Red blood cells ($\times 10^{12}$/l)	3.5	5
Reticulocytes (%)	12	0.5–1.5
Bilirubin (μmol/l)	340	2–14
Red cell glucose-6-phosphate dehydrogenase	not detected	

Fig. 12. Patient's red cells, showing 'Heinz' bodies. (*Courtesy of Dr J. E. MacIver, Manchester Royal Infirmary*)

examination, were found to contain small dark inclusion bodies ('Heinz bodies') (Fig. 12).

Five days later, the urine colour began to return to normal and the patient, still on primaquine, was feeling better. The haemoglobin concentration and the red cell count rose and the reticulocyte count declined. After a further few days the blood picture was normal. The red cell glucose-6-phosphate dehydrogenase level was 8 per cent of the normal value. The symptoms vanished and the patient was discharged. The diagnosis was malaria and primaquine sensitivity due to glucose-6-phosphate dehydrogenase deficiency.

Discussion

The administration of primaquine has evidently led to lysis of the patient's red cells. Much haemoglobin is liberated, some of which finds its way into the urine and colours it black ('blackwater fever'). The bulk of it is metabolized in the reticulo-endothelial system: the iron and the globin are removed and re-used, while the porphyrin ring is opened and converted to bilirubin, a water-insoluble yellow pigment. On account of its insolubility it is carried in the bloodstream attached to albumin and, when present in excess, is deposited in many tissues, colouring them yellow, e.g. this patient's sclerae. The condition is referred to as jaundice. Normally the slow formation of bilirubin precludes deposition in the tissues and enables it to be carried to the liver, where it is conjugated with glucuronic acid to render it water-soluble and non-toxic. The bilirubin glucuronide is excreted in the bile and is converted into urobilinogen in the gut, from where a portion of it is reabsorbed to be re-excreted by the liver and the kidney (Fig. 13).

The massive haemolysis engenders premature release of reticulocytes from the marrow, which accounts for the pronounced reticulocytosis observed in the patient.

Primaquine is one of many drugs (e.g. aspirin, sulphonamides) which are fairly innocuous to most patients but which can have disastrous effects on a few: individuals who have an inherited deficiency of glucose-6-phosphate dehydrogenase.

The function of this enzyme in the red cell is to generate reducing power, first in the form of NADPH and secondarily of glutathione (Fig. 14a and b). This tripeptide, composed of glutamic acid, cysteine (with its -SH group) and glycine, probably serves to keep -SH groups of membrane and other cellular proteins, including haemoglobin, in the reduced state (Fig. 14c).

In ordinary circumstances the glucose-6-phosphate dehydrogenase in the patient's cells (although none was detected in the first test) is barely adequate to generate the NADPH for the cell's needs. Most of the protein -SH is maintained in the reduced state, but some haemoglobin is

Fig. 13. Metabolism and disposal of haemoglobin

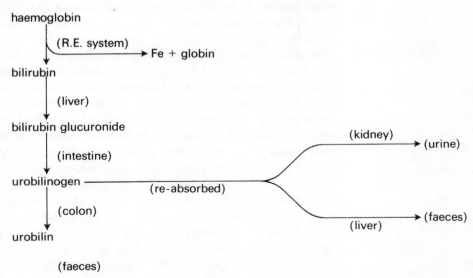

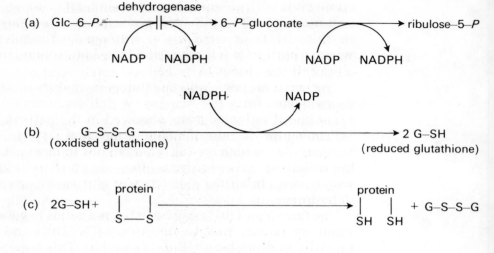

Fig. 14. Generation and use of reducing power in red cells

oxidized and yields degradation products, which precipitate within the cell to form the Heinz bodies. Primaquine and other drugs exacerbate the deficiency by tending to oxidize glutathione (G-SH) to G-S-S-G (Fig. 14d), thus making additional demands on the reducing-power-generating system. It is in the presence of such drugs that the deficiency of glucose-6-phosphate dehydrogenase becomes critical. Young red cells, with their intact stock of enzymes synthesized before the loss of the nucleus and the ribosomes, are just able to cope with the reduction of primaquine in addition to that of protein disulphides and methaemoglobin. Older cells, however, which have lost much of their original enzyme complement, including the deficient glucose-6-phosphate dehydrogenase, and cannot replenish it, are unable to generate the required amount of NADPH. When they are exposed to the drug, oxidation of their cellular protein is inescapable; haemoglobin is converted into a peroxide and then to methaemoglobin and may even form S-S bridges with membrane -SH. It is not too fanciful to suppose that such a linking of haemoglobin to the membrane or, alternatively, oxidation of membrane lipids, could alter the stability of the red cell and so doom it to haemolysis or destruction by the spleen.

Thus administration of primaquine leads to rapid

haemolysis of the patient's old red cells, which are metabolically incompetent to deal with the extra demand on their depleted reducing power potential. They are quickly replaced by young cells which contain about 8 per cent of the normal dehydrogenase activity and so ensure the patient's recovery, despite the continued presence of the drug.

The genetics of glucose-6-phosphate dehydrogenase deficiency poses some interesting questions. First, since the enzyme abnormality is caused by a defective gene, would one not expect a reduced enzyme activity in other tissues, as well as in the red cell? In fact, the enzyme defect is present in other tissues, but it is less severe. Possibly the enzyme is not very stable and in the absence of continuous synthesis the activity decays in the red cells very quickly. The health of the lens, like that of erythrocytes, depends greatly on the maintenance of reducing power and, hence, it also often suffers from the inherited defect; cataracts are the result.

Second, the disease is sex-linked, i.e. the gene for glucose-6-phosphate dehydrogenase is on the X chromosome. Since females have two X chromosomes and males only one, a normal female would be expected to have twice as much enzyme as a normal male, but this turns out not to be so. Both sexes have roughly the same amount of enzyme and hence it seems that in females only one gene is active in any one somatic cell. A heterozygous female produces two types of red cells: one containing the normal dehydrogenase, the other the defective enzyme. Males cannot be heterozygous for the condition, only having a single gene instead of the usual pair, and hence they are either normal or they have the full-blown deficiency.

Finally, the geographical distribution of glucose-6-phosphate dehydrogenase deficiency in the malarial regions of Africa, Asia and around the Mediterranean, suggests that the gene may confer some protection against malaria, thus ensuring its survival. Quite probably the malaria parasite relies on the pentose phosphate cycle and on the glutathione of the host erythrocyte for optimal growth, which may thus be impaired in the glucose-6-phosphate-dehydrogenase-deficient cell, so making malarial infection less severe in children with this inherited defect.

Questions

1. It is suggested above that the red cells could lyse as a consequence of a reduced membrane stability. Can you suggest other possible causes of haemolysis in this condition?

2. What are the functions of glucose-6-phosphate dehydrogenase in other tissues?

3. Unconjugated bilirubin is not excreted in the urine. Explain.

Alcoholism with Gout

Case History

J.C., a bank manager aged 46, was brought to the hospital after collapsing at a business meeting. On admission he was in a state of inebriation. His plasma alcohol concentration was 82 mmol/l, glucose was low, and the lactate and urate levels were raised (Table 10). At 2 a.m. the following morning he had an attack of excruciating pain in

Table 10. Concentration of alcohol and other plasma components

Component	Patient	Normal range
	mmol/l	
Alcohol	82	17.4*
Glucose	3.1	3.3–8.4
Lactate	2.8	0.7–2.0
Urate	0.6	0.2–0.5

*Legal limit for motorcar drivers in Great Britain.

his big toe. Later that morning the plasma lactate had returned to within the normal range, but the urate was still elevated at 0.54 mmol/l.

On questioning, J.C. reported that he had had spasmodic pain in his toes for several months and he admitted that he had been a heavy drinker for some time. Nodules ('tophi')

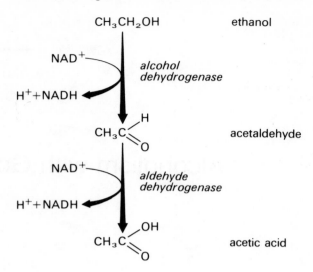

could be seen and felt in his earlobes—a characteristic feature of gout (Plate 4)—and the joints of his big toe were inflamed and exquisitely tender. He was given allopurinol and when, 3 days later, the plasma urate had fallen to normal levels, he was discharged. The diagnosis was gouty arthritis exacerbated by chronic alcoholism.

Discussion

Alcohol is metabolized, mainly in the liver, in two stages, catalysed by alcohol and aldehyde dehydrogenases, with NAD as a hydrogen acceptor (Fig. 15). The presence of alcohol within the cell makes a heavy demand on a limited supply of NAD and consequently the NAD/NADH ratio falls. Other reactions depending on NAD will thus be curtailed (Fig. 16). In particular, the oxidation of lactate to pyruvate will become much slower than the reverse reaction, which accounts for the observed accumulation of lactate.

Other redox reactions depending on NAD occur in the Krebs cycle, and these, too, are similarly slowed in the presence of alcohol, so that the maximum rate of oxidation of acetate may be not more than 25 per cent of the normal. Accordingly, acetate is diverted into the fatty acid synthetase system so accounting for the fatty liver commonly found in alcoholics.

The hypoglycaemia is the result of the failure of gluconeogenesis: the production of glucose from lactate,

Fig. 16. Effect of alcohol on pyruvate metabolism. The rates of reactions marked ‖ are reduced as a result of lack of NAD

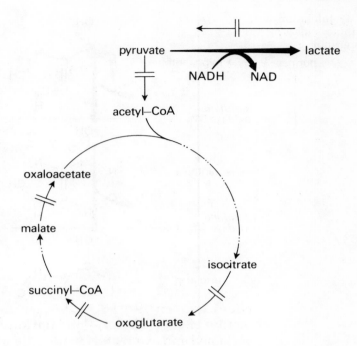

glycerol and amino acids. All of these precursors require NAD for the synthesis of glucose and since the oxidised form of the coenzyme is in short supply, gluconeogenesis is greatly impeded. Alcoholics often lose interest in food and hence their glycogen reserves may be so low that, with the failure of gluconeogenesis after indulging in alcohol, they can pass directly from alcoholic stupor to severe hypoglycaemia and die in coma.

The relationship between alcohol and urate needs to be explained. Lactate, when present in abnormally high concentration in the blood, competes in the renal mechanism for excreting urate. The kidney is less able to excrete this end product of purine metabolism and hence its plasma concentration will rise.

For some reason, as yet incompletely understood, the monosodium salt of uric acid, the form in which it is present in the plasma, tends to crystallize in cartilagenous tissue such as the ear lobes to form 'tophi' (Plate 4) and in the joint cavities causing the intense pain, both characteristic of gout. The big toe receives a disproportionate share of urate crystals and of the miseries they inflict, possibly due to degenerative disease being more advanced in this weight-bearing joint than in others.

Fig. 17. Inhibition of
xanthine oxidase by
allopurinol

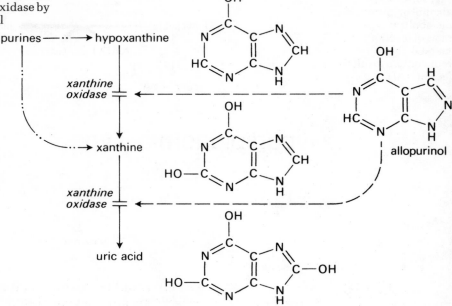

Allopurinol inhibits xanthine oxidase, an enzyme
which catalyses the oxidation of hypoxanthine and xanthine to uric acid. The drug is similar in structure to that of
uric acid (Fig. 17) and inhibits the oxidase competitively.
Xanthine is much more soluble than uric acid and the
kidney is therefore able to excrete this purine metabolite
without any precipitation occurring in the joints or
elsewhere.

Further Reading

Madison, L. L. (1968) *Adv. Metab. Disord.* 3, 85.

Questions

1. What is the origin of the lactic acid in the plasma?

2. Fructose speeds up the metabolism of ethanol to a
limited extent. Can you explain the mechanism?

Hyperlipidaemia Type I

Case History Lucy N., aged 13, was admitted to hospital at 9 a.m. with suspected acute appendicitis after severe abdominal pain throughout the night. On examination her liver and spleen were enlarged and numerous yellow patches (xanthomas) were seen on her trunk and limbs (Plate 6). A long history of abdominal pain had been ascribed to a 'grumbling appendix'.

A blood specimen had the appearance of 'creamed tomato soup' and after centrifugation a thick band of 'cream' collected at the top (Plate 7). Lucy had had nothing to eat for 15 h and the presence of much fat in the blood so long after the last meal suggested an abnormality of fat metabolism or transport. Determination of the plasma lipids confirmed hyperlipidaemia, the lipids consisting largely of triglycerides; the cholesterol was also raised (Table 11). Electrophoresis of the plasma lipoproteins confirmed the presence of chylomicrons (Fig. 18) and

Table 11. Plasma lipids (fasting)

Lipids	Patient	Normal range
	mmol/l	
Triglycerides	40	0.1–1.5
Cholesterol	13	3.5–5.5

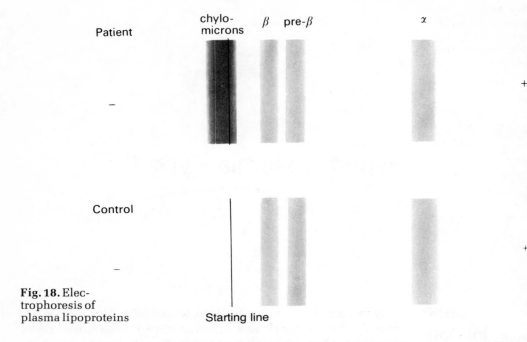

Fig. 18. Electrophoresis of plasma lipoproteins

enabled a diagnosis of hyperchylomicronaemia or hyperlipidaemia type I to be made.

Discussion

In order to confirm the diagnosis, the lipoprotein lipase activity of the plasma was measured at intervals after intravenous administration of heparin. In contrast to the normal response of a control subject, little fatty acid appeared in the patient's plasma (Fig. 19).

Lucy was prescribed a low-fat diet, with ample protein and carbohydrate to provide the necessary energy. The hyperlipidaemia regressed, the abdominal pain disappeared, and after a few weeks the xanthomas vanished. Her subsequent development was reasonably normal, although it was punctuated by episodes of violent abdominal pain, hyperlipidaemia and reappearance of the xanthomatous plaques, seemingly related to excessive consumption of fat.

This type of hyperlipidaemia is due to an hereditary deficiency of lipoprotein lipase, an enzyme which plays a vital role in the metabolism of triglycerides. Digestion of dietary fat by pancreatic lipase and absorption of fatty acids is followed by re-assembly, within the intestinal mucosa, of triglycerides which coalesce into droplets

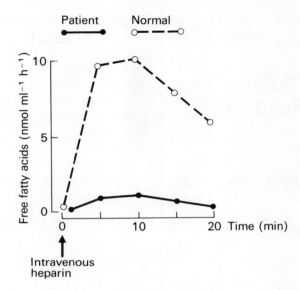

Fig. 19. Lipoprotein lipase activity in serum after administration of heparin

(chylomicrons), stabilized by a surface layer of lipoprotein, cholesterol and phospholipid. The amount of these droplets formed is clearly dependent on the fat intake and a raised blood level is found in normal subjects after a fatty meal. They are soon removed, however, by the adipose tissue and by metabolism in skeletal muscle and other tissues, a process preceded by hydrolysis to glycerol or lower glycerides and free fatty acids, catalysed by lipoprotein lipase. The enzyme is present in the cell membrane of many tissues and especially in capillary endothelium. It differs from pancreatic lipase in acting on lipoprotein-bound rather than free triglyceride. Lipoprotein lipase is not demonstrable in the blood of fasting subjects but it appears after ingestion of fat or intravenous administration of heparin. This polyanion probably liberates the enzyme from a combination with tissue protein and enables it to pass into the bloodstream.

The patient's plasma enzyme rose, after heparin, to not more than 10–15 per cent of the normal level. Whether this is due to a failure of the tissues to synthesize adequate amounts of the lipase or whether the enzyme is less active than in healthy subjects, remains to be determined.

Hyperlipidaemia is undoubtedly due to the failure to remove chylomicrons from the blood, secondary to deficient lipoprotein lipase activity in the tissues. Accumulation of triglycerides, cholesterol and phospholipids in the dermis gives rise to the xanthomatous plaques. The precise cause of the abdominal pain is unknown.

Further Reading

La Rosa, J. C., Levy, R. I., Windmueller, H. G. & Frederickson, D. S. (1972) Comparison of the triglyceride lipase of liver, adipose tissue and postheparin plasma, *J. Lipid Res.* 13, 356.

Questions

1. How are medium-chain triglycerides absorbed? Would they help in the treatment of this disease?

2. Why was the plasma cholesterol raised?

Myocardial Infarction

Case History

S.P., a man aged 58, was admitted to hospital for an exploratory chest operation. While surgery was in progress he suffered a coronary thrombosis, which led to cardiac arrest. Massage and resuscitation failed to restart his heart and he died.

A piece of heart tissue was taken from the area of the infarct and another from an unaffected part, and both were frozen immediately. Analysis of the two pieces of tissue carried out later showed that the lactate concentration had increased tenfold, while the immediate sources of cell energy, ATP, creatine phosphate and glycogen, had decreased considerably in the infarcted compared to the normal tissue (Table 12).

Table 12. Metabolites in normal and infarcted tissue

Metabolite	Normal	Infarcted
Glycogen (mg/g)	4.1	1.9
Lactate (μmol/g)	2.1	20
ATP (μmol/g)	5.5	1.3
Creatine phosphate (μmol/g)	8.0	1.5

Discussion

Thrombosis of a coronary artery has led to an interruption of the blood supply to an area of the heart, with disastrous consequences. Oxidative phosphorylation stopped and anaerobic glycolysis was accelerated. Despite the increased glycogenolysis the ATP concentration dropped rapidly to one-quarter of the normal value, the reserve of creatine phosphate being almost depleted by the action of creatine kinase. Lactate accumulated owing to the failure of oxidation of NADH in the electron transport chain (Fig. 20).

Fig. 20. Generation of ATP in infarcted tissue

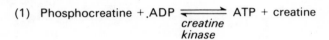

(1) Phosphocreatine + .ADP $\rightleftharpoons$ ATP + creatine
creatine kinase

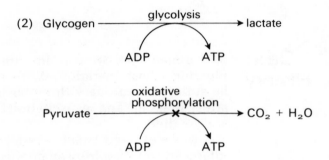

(2) Glycogen ——— glycolysis ———→ lactate
ADP   ATP

Pyruvate ——— oxidative phosphorylation ——✗——→ $CO_2 + H_2O$
ADP   ATP

The drastic fall in ATP would be expected to result in early failure of the sodium pump, which consumes a substantial proportion of the cell's energy, and hence in changes in the cellular cation concentrations. Sodium ions will diffuse into the cells, impelled both by the concentration gradient and by the electrical potential, whereas potassium ions will diffuse outwards down the concentration gradient but against the potential. The total cellular cation concentration will therefore rise and attract water. The tissue becomes oedematous, the mitochondria swell, their structure becomes disorganized and the cristae collapse. In the prevailing conditions lysosomal enzymes are released and begin the process of degradation upon which the autolysis of dead cells depends.

If ischaemia is severe, the biochemical changes lead to irreversible damage and ultimately to death of the cells throughout the affected tissue. The damage is the greater and the more rapid, the greater the oxygen requirements of the tissue concerned; tissues depending on high metabolic activity are least able to withstand lack of oxygen and other nutrients. Cells of metabolically less demanding tissues or those on the periphery of an ischaemic lesion can often survive as a result of the expansion of collateral channels.

Thrombosis may be a consequence of an exaggeration or a perversion of the early stages of the normal blood clotting mechanism. These involve, first, an interaction of blood platelets with the amino groups of collagen, which results in release of ADP and some other constituents from the platelets. More platelets then adhere to the first layer, a process depending on an increase in 'adhesiveness' engendered specifically by ADP. The precise mechanism of this action is not understood, but the nucleotide is not itself incorporated into the platelet aggregate, suggesting that it does not act as 'glue'. Subsequently the platelet mass becomes tightly packed, probably as a result of activation of a contractile mechanism involving an actomyosin-like protein within the platelets.

Intravascular platelet adhesion and aggregation can occur without activation of the coagulation mechanism which comes into action when blood is shed; but it seems that the vascular endothelium must have suffered some as yet undefined and possibly quite subtle damage before the platelets can attach to it and cause thrombus formation. Deposition of atherosclerotic plaques may be a major factor in thrombogenesis, by damaging the vascular endothelium and by slowing down the blood flow which would otherwise sweep away any incipient platelet aggregates.

Prostaglandin E_1 and its mediator, cyclic AMP (adenosine 3′:5′-monophosphate), appear to play a role in the regulation of platelet behaviour: they inhibit platelet aggregation *in vitro*, possibly by reducing adhesiveness. Both phenomena are direct consequences of membrane surface structure. There is some evidence that a cyclic-AMP-activated protein kinase may phosphorylate a protein within the platelet membrane and it is not unreasonable to suppose that the introduction of phosphate groups could so alter the surface properties as to reduce adhesion between platelets.

Spontaneous aggregation *in vivo* is associated with a

decrease in cyclic AMP concentration within the platelets, a fact which might well yield clues to the problem of thrombus formation.

Questions

1. By what mechanisms is glycogenolysis stimulated in hypoxic tissue?

2. What histological changes would you see in the infarcted tissue?

Ketoacidosis in Diabetes

Case History Ann P., a housewife aged 35, was brought to hospital in semicoma. She had been treated for diabetes for the past 12 years. On examination she showed signs of dehydration: the mucous membranes were dry, the skin was inelastic and wrinkled and the eyeballs were sunken. Breathing was deep and rapid and the breath had the odour of acetone. Blood pressure was low and the pulse fast. The patient was able to pass a specimen of urine which gave strongly positive tests for glucose and ketone bodies. A blood sample was taken and a rapid semi-quantitative test for glucose indicated a level in excess of 15 mmol/l.

While the blood was sent to the laboratory for detailed analysis, the patient was given an intravenous drip of physiological salt solution and insulin. The laboratory

Table 13. Patient's plasma: laboratory data

Plasma component	Patient	Normal range
HCO_3^- (mmol/l)	12	24–35
pH	7.1	7.3–7.5
Urea (mmol/l)	8	2.5–7.5
Osmolality (mosmol/l)	385	285–295
Na^+ (mmol/l)	136	138–150
K^+ (mmol/l)	5.8	3.9–5.6
Ketone bodies (mg/dl)	350	0–3

investigation yielded the information presented in Table 13. The urine contained microorganisms later identified as *Escherichia coli*. For this infection of her urinary tract she was treated with antibiotics.

Four hours later the patient's clinical condition was much improved. She was given more insulin and the composition of the saline drip was adjusted in accordance with the changes in her blood electrolytes. The treatment was continued and within 2 days the signs of dehydration and of ketoacidosis had disappeared. The patient could be stabilized on her former dose of insulin and was discharged.

Discussion

This patient's diabetes had been well controlled by regular administration of insulin, but the stress caused by the urinary tract infection had, presumably, upset the balance of endogenous hormones and administered insulin. Since glucose transport across cell membranes in muscle and some other tissues depends on insulin, the temporary deficiency of the hormone had reduced the amount of hexose available to these tissues.

The dual threat of energy and glucose deficiency is met in two ways: glucocorticoids are secreted by the adrenal cortex to stimulate gluconeogenesis, and growth hormone secretion accelerates lipolysis in adipose tissue to provide fatty acids for oxidation. In response to glucocorticoids tissue proteins are broken down into amino acids: some yield pyruvate or α-oxoglutarate, while others, the 'ketogenic' amino acids, e.g. leucine, can form acetyl-CoA or acetoacetyl-CoA. A major contribution to the supply of the carbon skeleton in gluconeogenesis is probably made by alanine, which yields first pyruvate and then oxaloacetate, phospho*enol*pyruvate and ultimately glucose. Glutamate and glycerol are also substrates (Fig. 21).

Hyperglycaemia, which results from inability of muscle to take up plasma glucose in the absence of insulin, is thus further aggravated by gluconeogenesis.

Stimulation of lipolysis and β-oxidation of fatty acids results in formation of acetyl-CoA, to be oxidized to CO_2 and H_2O in the Krebs cycle and electron transport chain and thus to yield ATP. It is possible that there is not enough oxaloacetate in the mitochondria for efficient running of the cycle, especially in view of the generation, in the course of fatty acid oxidation, of much NADH which

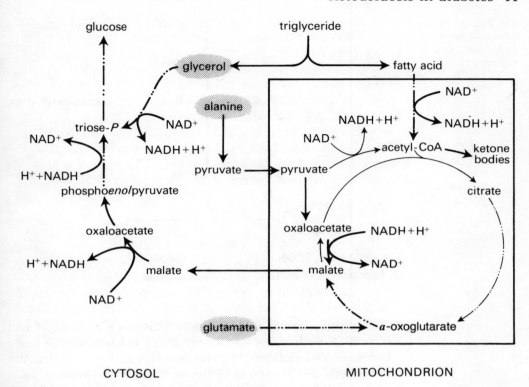

Fig. 21.
Gluconeogenesis
and lipolysis in
uncontrolled dia-
betes. Major sub-
strates for
gluconeogenesis in
shaded area

would tend to reverse the malic dehydrogenase reaction
towards malate (Fig. 21). Alternatively it could be
assumed that the Krebs cycle is working at maximum
efficiency but cannot cope with an excessive amount of
acetyl-CoA, which is, therefore, diverted to acetoacetyl-
CoA and to free acetoacetate (Fig. 22). The keto acid can be
further metabolized by reduction or decarboxylation to
β-hydroxybutyric acid and acetone respectively. The three
substances are known collectively as ketone bodies. They
accumulate in the patient's blood (Table 13) as a result of
the increased synthesis. Acetone, with its high vapour
pressure, is exhaled in the lungs and manifests itself in the
breath.

Acetoacetic and hydroxybutyric acids are moderately
strong acids, which lower the pH of the blood and of the
urine. At the increased acidity of the blood the bicarbonate
concentration falls and more carbonic acid is formed (Fig.
23), which stimulates the respiratory centre to remove
excess CO_2, so accounting for the patient's rapid, deep
respiration.

Fig. 22. Generation of ketone bodies from acetyl-CoA in diabetes

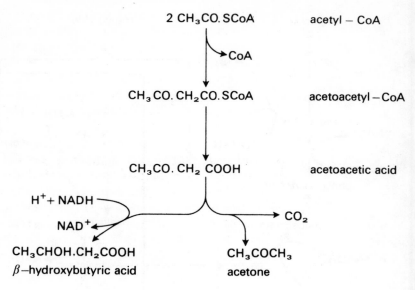

$2 CH_3CO.SCoA$ acetyl $-$ CoA

CoA

$CH_3CO.CH_2CO.SCoA$ acetoacetyl $-$ CoA

$CH_3CO.CH_2 COOH$ acetoacetic acid

$H^+ + NADH$

NAD^+

CO_2

$CH_3CHOH.CH_2COOH$ CH_3COCH_3

β–hydroxybutyric acid   acetone

Hyperglycaemia and glucosuria induce a prolonged and severe osmotic diuresis: much water is lost and the body becomes dehydrated. As the blood volume shrinks, the peripheral resistance is reduced, the blood pressure falls and the pulse accelerates. In these circumstances kidney function is impaired, hampering the excretion of glucose, ketone bodies, urea, etc. Thus, the extra-cellular fluid becomes hyperosmolar, so withdrawing water from the intracellular compartment. As Na^+ is lost in the urine to neutralize the excreted organic acids, K^+ and Mg^{2+} pass

Fig. 23. Effect of organic acids on plasma bicarbonate and pH

Normal equilibrium:

$$CO_2 + H_2O \rightleftharpoons H_2CO_3 \rightleftharpoons H^+ + HCO_3^-$$

Addition of H^+ derived from organic acids:

$$CO_2 + H_2O \longleftarrow H_2CO_3 \longleftarrow H^+ + HCO_3^-$$

Since $pH = pK + \log_{10} \dfrac{[HCO_3^-]}{[H_2CO_3]}$

and $[HCO_3^-]$ decreases more than $[H_2CO_3]$

$\therefore$ pH falls

out of the cells and are excreted. While a potassium deficiency develops in the cellular compartment, the plasma concentration of the ion shows a paradoxical rise (Table 13).

The combined effects of dehydration and acidosis, with consequent deficiency of water and electrolytes, greatly impair cerebral metabolism and function, leading eventually to coma.

Upon administration of insulin the growth hormone and glucocorticoid secretion is switched off, gluconeogenesis subsides and metabolism of ketone bodies accelerates. The intravenous drip slowly replenishes the body water and restores both extracellular and intracellular electrolytes. The blood pH returns to normal. Control of the patient's diabetes can be re-established by the appropriate daily dose of insulin.

Questions

1. The oxidation of malate by NAD, catalysed by malic dehydrogenase, to oxaloacetate is a highly endergonic reaction ($\Delta G^0 = +7$ kcal/mol; $+29$ kJ/mol). How can it go in the forward direction under normal physiological conditions?

2. In view of the Na^+ loss in the urine, why is the Na^+ concentration in the plasma almost normal?

Chinese Restaurant Syndrome

Case History
A party of four couples went to a Chinese restaurant and started their meal with Won-Ton soup, followed by a variety of Chinese dishes. Twenty minutes later three members of the party complained of a sudden tightening of the face, associated with numbness spreading from the jaw to the back of the neck, dizziness, flushing and sweating; and a severe pain in the neck and shoulders. Two people also complained of a band-like headache, palpitation and weakness. The other five members of the party were not affected.

All reported to the local hospital, but they had recovered within 30 min of the attack. A diagnosis of 'Chinese Restaurant Syndrome' was made.

Discussion
The offending component of the food is monosodium glutamate, widely used as an additive and not only in Chinese food. Manufacturers advocate as much as 1 g per serving 'to bring out the full "natural" flavour'. There is a dose-response relationship between monosodium glutamate and facial pressure, but the other complaints are more variable. The symptoms appear only if the meal is taken on an empty stomach by a susceptible individual. Free glutamic acid is similarly active. The acid occurs widely in the body, especially in the central nervous system where it is present in greater concentration than

Table 14. Free amino acids in human brain

Amino acid	μmol/g fresh tissue
Glutamate	10.6
Glutamine	4.3
γ-Aminobutyrate	2.3
Other amino acids	0.05–2.2

any other amino acid (Table 14). Glutamic decarboxylase, the enzyme which catalyses the decarboxylation of glutamate to γ-aminobutyrate, is found almost exclusively in nervous tissue (Fig. 24).

Glutamic acid is an excitatory amino acid, while γ-aminobutyric acid has an inhibitory effect on the electrical behaviour of the nervous system. In fact, the properties of the two acids make them candidates for natural transmitters. Although experiments show that injected glutamate has little or no effect on the glutamate concentration of the brain, an exchange between administered and endogenous amino acid can be demonstrated if labelled glutamate is used. It is possible, therefore, that the action of the ingested monosodium glutamate is indirect and based on disturbing an equilibrium involving glutamate, glutamine, NH_4^+ and α-oxoglutarate, on which the proper functioning of the nervous system may depend.

Fig. 24. Metabolism of glutamic acid

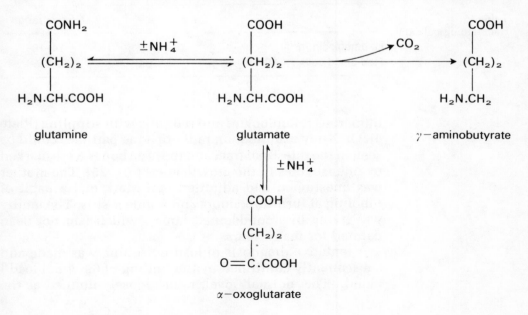

Lead Poisoning

Mark S., the 2-year-old child of an unskilled labourer, was referred to hospital for investigation after several weeks of vomiting, constipation, pallor and irritability. No abnormalities were apparent; the tendon reflexes were absent. Haematological examination revealed anaemia (Table 15) and a blood smear stained with methylene blue showed

Table 15.
Haematological data

	Patient	Normal range
Haemoglobin (g/dl)	7	14–18
Red blood cells ($\times 10^{12}$/l)	3.3	5

numerous reticulocytes and red cells with stippling (Plate 8). On X-ray examination radio-opaque particles could be seen in the intestinal tract and the long bones were marked by intense lines in the growth areas (Fig. 25). The mother was questioned and admitted that Mark had a habit of nibbling at furniture, doors and window sills. The family was living in a condemned house, which had not been painted for many years.

A tentative diagnosis of lead poisoning was made and subsequently confirmed by the finding of 6.8 μmol lead/l blood. (The 'normal' level, rather loosely defined as the

Fig. 25. X-ray of ankle joint: (A) patient, (B) control. Note bands of increased density at the growing areas of the metaphyses (arrows). (*Courtesy of the Department of Medical Illustration, Manchester Royal Infirmary*)

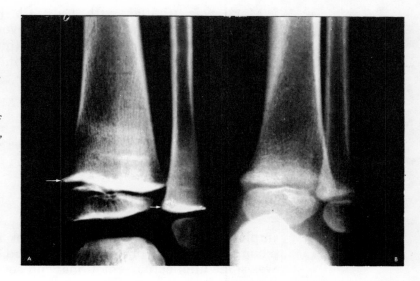

concentration above which clinical evidence of lead poisoning becomes apparent, is 2.4 μmol/l.) Qualitative tests on the urine indicated glucosuria and aminoaciduria, and determination of urinary porphyrin (coproporphyrin) and its metabolic precursor, δ-aminolaevulinic acid, showed excessive excretion of both (Table 16).

Table 16. Excretion of coproporphyrin and δ-aminolaevulinic acid in the urine

Urine component	Patient	Normal
	μmol/24 h	
Coproporphyrin	0.5	< 0.08
δ-Aminolaevulinic acid	30	trace

Fig. 26. Calcium EDTA. Lead displaces the Ca^{2+} and is excreted in the complexed state

Mark was treated with CaEDTA (calcium ethy-lenediaminetetraacetate) (Fig. 26) intravenously and after 24 h the daily excretion of lead had increased fiftyfold (Table 17). On the third day his condition began to

Table 17. Excretion of lead in urine

Sample	Patient	Normal
	μmol/24 h	
Before treatment	0.5	< 0.4
24 h after EDTA	25	—

improve and he was sent home a week later. When last examined at 4 years of age he was in good health, but there was some intellectual, emotional and social impairment.

Discussion

The urge to nibble paint and other objects is not uncommon in children. It is largely for this reason that the use of lead in paint is now illegal in many countries. The lead in Mark's body had clearly come from paint he nibbled off doors and windows in the family's old house and fragments were present in his intestinal tract on admission to hospital.

EDTA acts as a chelating agent, forming strong coordinate links with divalent and polyvalent cations to yield unionized complexes (Fig. 26). Other chelators, e.g. D-penicillamine and British anti-lewisite, act similarly and have been used for removing heavy metals from the body. Because of the relatively non-selective nature of cation binding, there is a danger of removing Ca^{2+} and hence the calcium complex of EDTA is employed for this therapy. In the conditions prevailing *in vivo* lead displaces calcium from the complex and so can be excreted by the kidney. In order to avoid dissolving the paint fragments in the patient's gut, the CaEDTA complex was given intravenously.

Lead ions are rapidly adsorbed on to red cells and soft tissues like brain, lungs and marrow by virtue of their great affinity for sulphydryl groups of proteins. The function of many proteins, enzymes and others, depends on the free -SH groups of their cysteine residues and when these combine with lead they become inactive. Lead is not

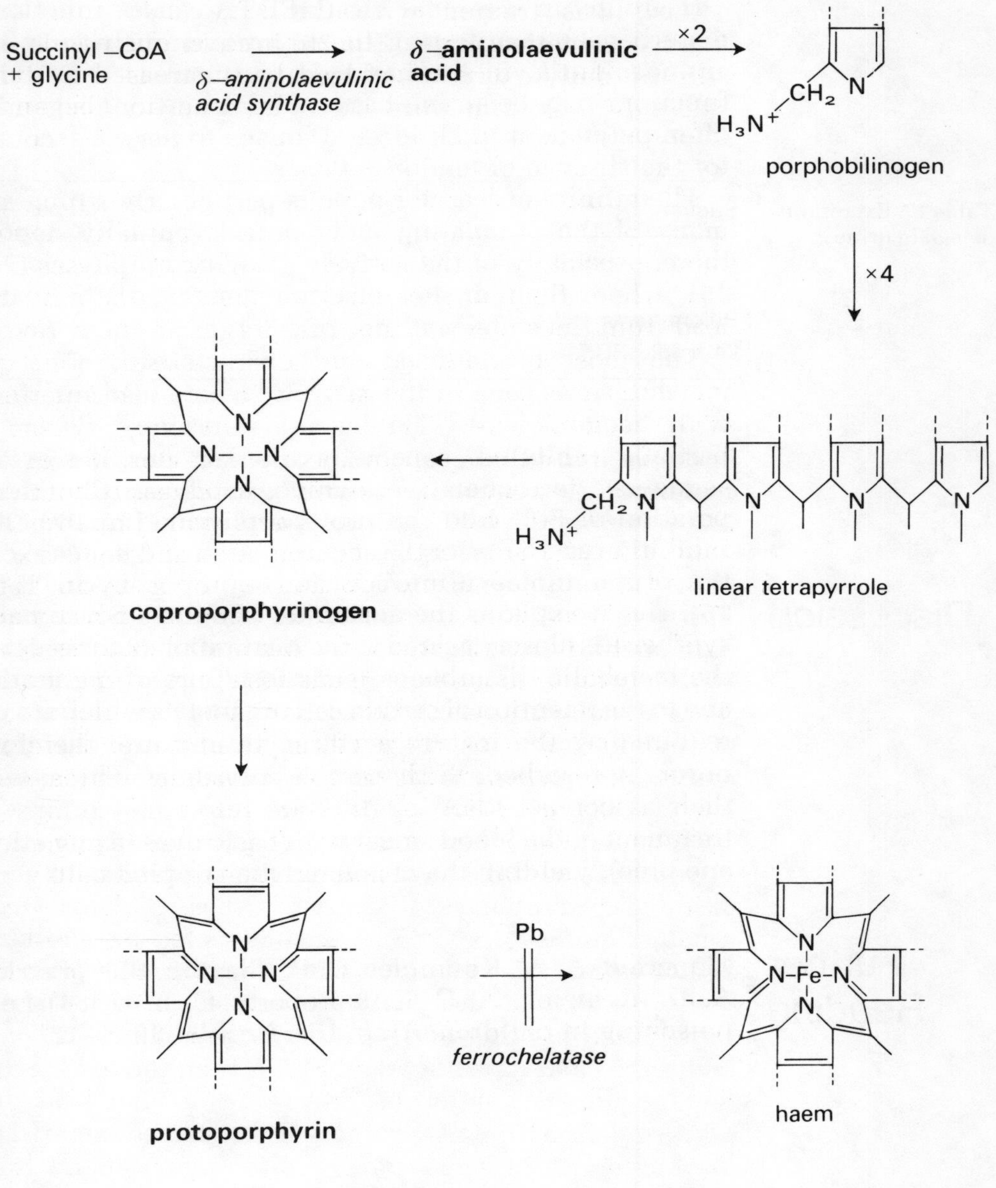

Fig. 27. Effect of lead on haem synthesis. Intermediates in bold type accumulate in the presence of lead

unique in the ability to inhibit enzymes, other heavy metals (e.g. cadmium, mercury, uranium) act similarly. The enzyme inhibition is of the non-competitive type, which cannot be relieved by raising the substrate concentration: those enzyme molecules which have lost their free -SH groups are completely inactive while the remaining molecules retain their full activity. Lead binding to

-SH groups in the renal tubular cells causes functional disturbances manifesting themselves in glucosuria and aminoaciduria. In the brain, impairment of the higher functions may be an indication of damage by lead and is often permanent in children. Damage to nerves accounts for the absence of tendon reflexes.

The affinity of lead for bone is particularly strong and many of the circulating metal ions eventually deposit there, especially in the actively growing epiphyses (Fig. 25), where their greater electron density distinguishes lead from the softer outlines of calcium on the X-ray.

The most pronounced and characteristic effect on metabolism is seen in the marrow, where lead interferes with haemopoiesis. The haem-synthesizing system is partially inhibited, especially the last enzyme of the sequence, ferrochelatase, which catalyses the incorporation of Fe^{2+} into the protoporphyrin (Fig. 27). The inhibition accounts for the accumulation and hence excretion of δ-aminolaevulinic acid and coproporphyrin (Table 16), and it explains the anaemia (Table 15). Since haem synthesis is closely geared to the maturation of the red cell, the metabolic disturbance leads to failure of maturation and to the retention of certain cell organelles which are not required by the mature erythrocyte and are, therefore, normally resorbed. Such are the polysomes which, with their associated RNA, agglutinate into small lumps on treatment of the blood smear with basic dyes (e.g. methylene blue), yielding the characteristic stippled cells.

Further Reading

Moncrieff, A. A., Koumides, O. P., Clayton, B. E., Patrick, A. D., Renwick, A. G. C. & Roberts, G. E. (1964) Lead poisoning in children, *Arch. Dis. Childh.* 39, 1–12.

Plate 1. Biopsy of jejunal mucosa, stained for lactase. Second specimen taken 6 weeks after admission. (*Courtesy of Prof. Z. Lojda, 1st Pathological Institute, University of Prague*)

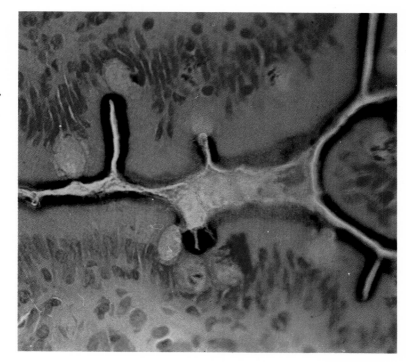

Plate 2. Liver biopsy: (A) normal, (B) patient with glycogen storage disease. Note extensive vacuolization at sites of intracellular glycogen deposition and consequent damage to liver cells. (*Courtesy of Dr S. Oleesky, Manchester Royal Infirmary*)

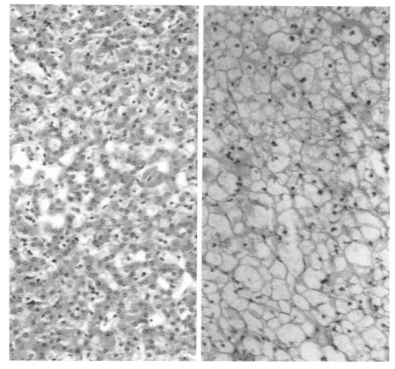

Plate 3. Liver biopsy, stained with periodic acid/Schiff for glycogen. (A) patient with glycogen storage disease; (B) normal; (C) patient, tissue incubated with amylase before staining. The large amount of glycogen present in the patient's liver is indicated by the intense pink stain (A), which has disappeared after incubation with amylase (C). (*Courtesy of Dr S. Oleesky, Manchester Royal Infirmary*)

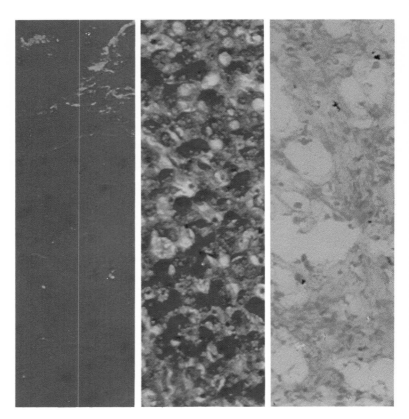

Plate 4. Tophi in patient's ear lobe. (*Courtesy of Department of Medical Illustration, Manchester Royal Infirmary*)

Plate 5. Patient's foot: note inflamed joints of big toe. (*Courtesy of Prof. J. H. Kellgren, Manchester Royal Infirmary*)

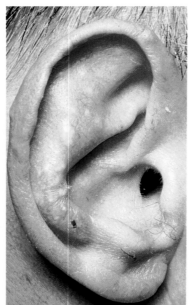

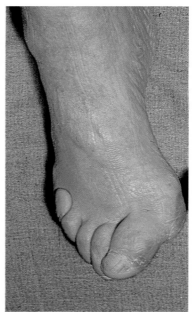

Plate 6. Patient's skin, showing xanthoma. (*Plates 6 and 7 are courtesy of Dr B. R. Tulloch, Manchester Royal Infirmary*)

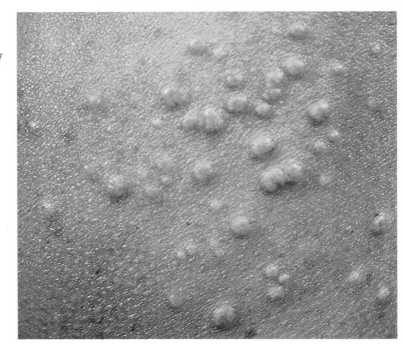

Plate 7. 'Cream' separated from patient's blood

Plate 8. Blood smear, stained with methylene blue, showing stippled cells. (*Courtesy Dr J. E. MacIver, Manchester Royal Infirmary*)

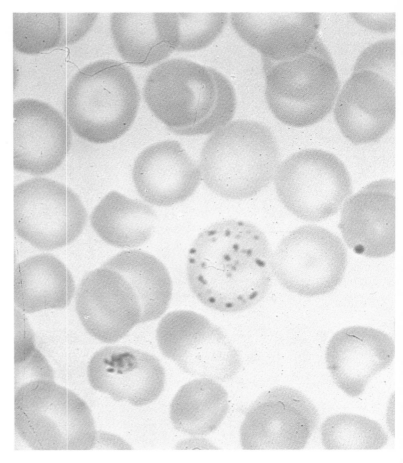

Plate 9. Patient John P. (*Courtesy of Dr I. B. Sardharwalla, Royal Manchester Children's Hospital*)

Plate 10. Patient's tongue, before treatment. (*Courtesy of the Department of Medical Illustration, Manchester Royal Infirmary*)

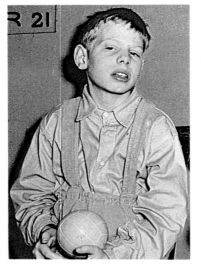

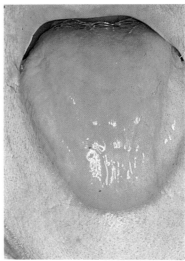

Phenylketonuria

Case History John P., aged 10, had been born after a normal pregnancy. He had taken his bottle feeds well for a few days, but then started vomiting frequently. Eventually vomiting ceased and John gained weight. He was slow in development, could not hold his head erect until 5 months and he could not sit unaided when he was 1 year old. The mother had observed a peculiar mousey odour of the boy's urine. On admission to hospital his urine was tested with ferric chloride: the characteristic green colour indicated the presence of phenylpyruvic acid. Quantitative analysis of the blood and urine yielded the values for phenylalanine and its metabolites shown in Table 18.

A diagnosis of phenylketonuria was made and John was put on a low-phenylalanine diet. After a few days the phenylalanine levels in the blood and urine were much lower and the phenylpyruvate and phenyllactate dis-

Table 18. Phenylalanine and its metabolites in blood and urine

Source	Metabolite	Patient	Normal
		mmol/l	
Plasma	phenylalanine	1.6	<0.09
Urine	phenylalanine	6.8	trace
	phenylpyruvate	4.5	0
	phenyllactate	10.2	0

appeared, as did the mousey odour of the urine (caused by another metabolite of phenylalanine).

John had started walking at 5½ years, he was incontinent, unable to feed himself and he could say only a few words. At age 10 the boy presented a picture of gross mental retardation, with an I.Q. of 65 (Plate 9). There were

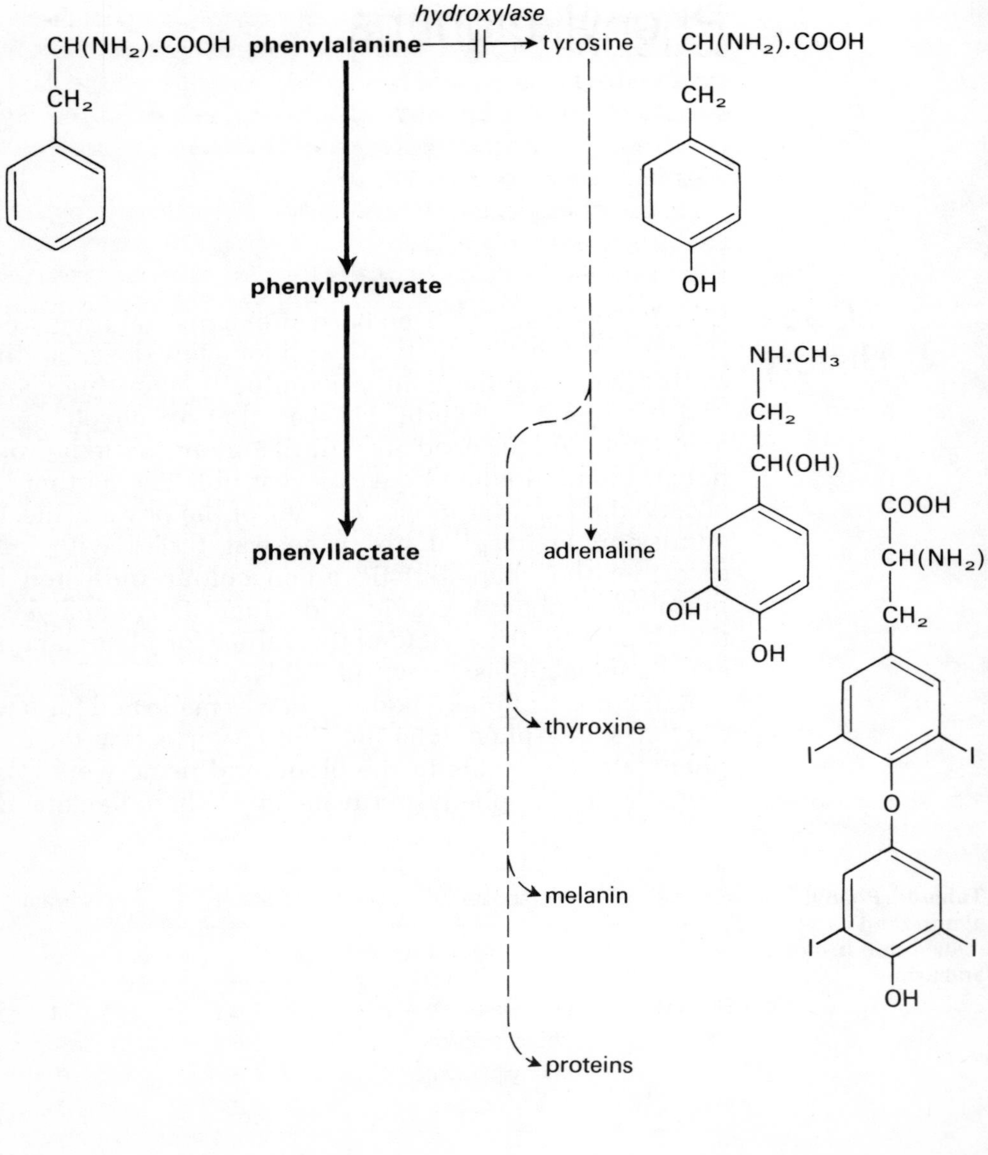

Fig. 28. Metabolism of phenylalanine; metabolic block in phenylketonuria

repetitive movements of his arms and legs; his hair was almost white in patches and the eyes were blue. He sat in his bed or walked about, often causing himself minor injuries by his uncontrolled movements.

Discussion   Phenylalanine accumulates in the blood and in the tissues, and is excreted in the urine, owing to a deficiency of phenylalanine hydroxylase in the liver. The enzyme catalyses the introduction of a hydroxyl group in the *para* position, yielding tyrosine. The patient cannot, therefore, synthesize tyrosine and has to rely entirely on the dietary supply of the amino acid, which is an essential precursor of adrenaline, noradrenaline and thyroxine, as well as the pigment melanin (Fig. 28).

As the phenylalanine level in the blood rises, metabolic pathways open up which are not normally functional. In this manner phenylpyruvate is formed, which gives rise to phenyllactate, both being found in the blood and urine of the patient. Apart from the potential deficiency of tyrosine, amino acid and protein metabolism may be deranged by the accumulation of phenylalanine, which may interfere with the selective transport of other amino acids across cell membranes. In the kidney, for instance, the tubular transport system faces a load of phenylalanine which approaches its maximum transport capacity, and in the brain the amino acid inhibits the transport of tyrosine, 5-hydroxytryptophan and other amino acids, leading to serious distortions of amino acid patterns in the cerebrospinal fluid and brain. The injurious effect of an imbalance of amino acids has been demonstrated by feeding phenylalanine to healthy experimental animals.

Phenylalanine is also believed to inhibit 5-hydroxytryptophan decarboxylase, an important enzyme which catalyses the production of 5-hydroxytryptamine (serotonin) (Fig. 29).

Fig. 29. Formation of 5-hydroxytryptamine (serotonin)

$decarboxylase$ CO_2

$CH_2CH(NH_2).COOH$? phenylalanine $CH_2CH_2NH_2$

5–hydroxytryptophan   5–hydroxytryptamine

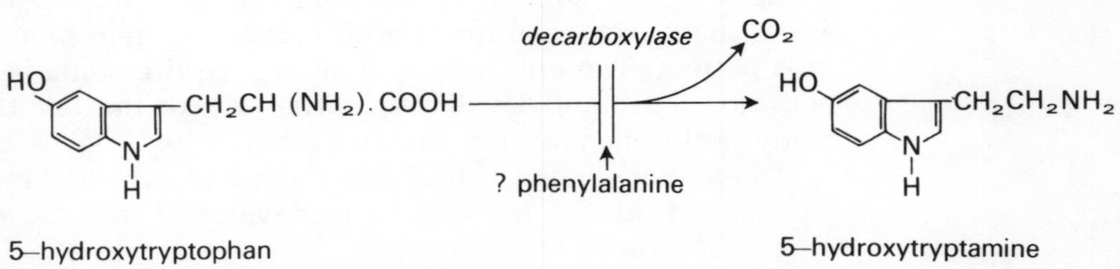

The precise manner in which the brain of the phenylketonuric patient is damaged is not known and there may well be a number of contributory factors, including the effect of amino acid imbalance on protein synthesis, interference with myelination and deficient production of serotonin, adrenaline and noradrenaline. The damage is irreversible, but the behavioral abnormalities appear to be due to raised levels of pharmacodynamic amines derived from phenylalanine and when these return to normal on a restricted phenylalanine diet, the behaviour disturbances disappear.

The light-coloured patches on the hair are due to defective melanin synthesis, which also accounts for lighter skin and eye pigmentation than might be expected genetically. Negro phenylketonurics with sandy hair, grey eyes and light skin have been described.

The low-phenylalanine diet is based on casein hydrolysates which have been passed through charcoal to remove aromatic amino acids, all of which are then replaced except phenylalanine. Early diagnosis and treatment with this diet prevent the intellectual impairment, but delay beyond the first few weeks of life results in permanent damage to the immature brain. In many areas screening programmes are in operation which aim at testing urine of every infant for phenylpyruvate by the simple $FeCl_3$ test. Any positive findings are then investigated further by determinations of the plasma level of phenylalanine and the appropriate dietary treatment is given to proven cases of phenylketonuria. In this manner brain damage can be prevented and the affected children develop normally despite their enzyme defect.

Very careful control of the diet is essential to ensure a supply of phenylalanine and tyrosine adequate for the body's requirements but not such as to cause phenylalaninaemia, which would interfere with the transport and metabolism of other amino acids. Too little phenylalanine results in serious malnutrition with potentially disastrous consequences, while too much leads to the gross disturbances of the central nervous system seen in this patient. The only way of gauging requirements is to monitor the blood level frequently—a task the healthy body performs continuously and efficiently.

Recent work suggests that after the age of 8 the diet may be relaxed, as the brain is fully developed and excess phenylalanine is then harmless.

Questions
1. How do you account for the opening up of metabolic pathways, not normally functional, when the blood level of phenylalanine rises?

2. In this patient the hair is white in patches; in others it is uniformly light in colour. What might be the reason for the patchy appearance?

Homocystinuria

Case History

Paul B., aged 8, was brought to hospital at his parents' request at the time his sister was under investigation for a presumed hereditary metabolic disorder. He was knock-kneed and his feet were abnormally arched (Fig. 30). He had fine, brittle, sparse hair and dislocated lenses (Fig. 31). His mental capacity was very low, he was neither clean nor dry and he could say only a few words.

Biochemical studies yielded the following information. The urine contained an increased level of amino acids, and on ion-exchange chromatography a peak was identified as homocystine, which does not occur in normal urine. The plasma also contained homocystine, as well as an unduly high concentration of methionine (Table 19). In view of

Table 19. Amino acid concentration in plasma

Amino Acid	Patient	Normal range
		μmol/l
Homocystine	56	0
Methionine	114	20 – 40

the elevated plasma methionine level a loading test was carried out: after an oral dose of methionine the excretion of this amino acid and of homocystine in the urine was

Fig. 30. Patient Paul B. (*Courtesy of Dr G. M. Komrower, Royal Manchester Children's Hospital*)

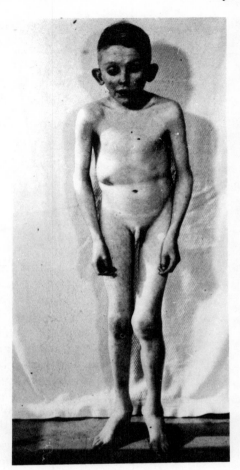

Fig. 31. Patient's eyes, showing the more severely dislocated lens. (*Courtesy of Dr G. M. Komrower, Royal Manchester Children's Hospital*)

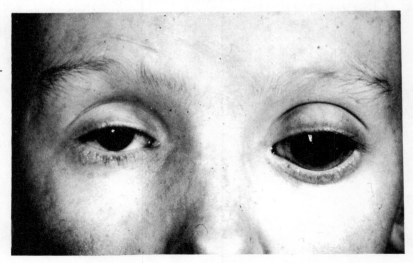

Fig. 32. Methionine loading test

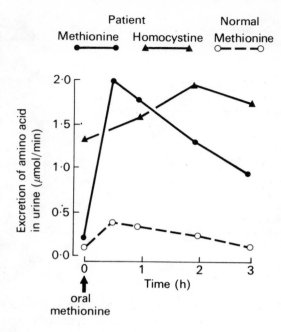

monitored for 3 h (Fig. 32). The patient excreted a much larger proportion of the administered methionine than did the controls and the homocystine in his urine rose. On the basis of these findings a diagnosis of homocystinuria was made.

Discussion

Homocysteine is chemically a close relative of cysteine, the only difference being a -CH$_2$ group (Fig. 33). The two amino acids are closely related metabolically as well: homocysteine condenses with serine to form cysta-thionine, which is then split up into homoserine and

Fig. 33. Structures of cysteine and homocysteine

CH(NH$_2$)COOH
|
CH$_2$
|
SH

Cysteine

CH(NH$_2$).COOH
|
CH$_2$
|
CH$_2$
|
SH

Homocysteine

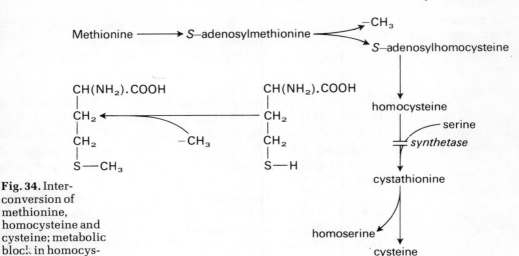

Fig. 34. Interconversion of methionine, homocysteine and cysteine; metabolic block in homocystinuria

cysteine (Fig. 34). The first stage of the process is catalysed by cystathionine synthetase, an enzyme which is not functional in this patient. The defect is inherited. (The case history suggests that both siblings may be affected in this family.)

The metabolic block proximal to cystathionine results in accumulation of homocysteine and the disulphide homocystine (analogous to cystine) which is found in the blood and urine. Another consequence of the presence in the tissues of an abnormally large amount of homocysteine is its mass action effect on the reaction with methyl donors to re-form methionine (Fig. 34). Thus the enzymatic block in the metabolic sequence connecting methionine and cysteine accounts for the accumulation of methionine as well as homocystine.

How can the triad of biochemical abnormalities —accumulation of homocystine and methionine and a potential deficiency of cysteine, stemming directly from the genetic defect—explain the pathogenesis of homocystinuria? The structural similarity of the three sulphur-containing amino acids might lead one to suspect that the aminoacyl-tRNA synthetases, which catalyse the coupling of amino acids to their specific tRNAs, could be 'deceived' by homologues and so lead to their incorporation into proteins, producing metabolic or structural misfits. It has long been known that ethionine, if present at high concentration, is coupled to the tRNA specific for methionine.

Fig. 35. Formation of homocysteine transfer RNA

$$\text{Cysteine} + \text{tRNA}^{Cys} \xrightarrow{\hspace{2cm}} \text{Cys}-\text{tRNA}^{Cys}$$

synthetase

$$\text{Homocysteine} + \text{tRNA}^{Cys} - - - - \rightarrow \text{Hcy}-\text{tRNA}^{Cys}$$

Similarly homocysteine seems capable of being attached to tRNA^{Cys} (Fig. 35) and thus finding its way into proteins. Hair, with its normally high cystine content, is an obvious candidate for such an error of synthesis, and indeed a small amount of homocysteine was found in the patient's hair.

If the sparsity and the abnormal physical state of the hair cannot be ascribed entirely to a qualitative fault, perhaps they are due to a quantitatively defective synthesis arising from a deficiency of cysteine. A moderate lack of specificity of the carrier which transports cysteine across the cell membrane into the cytoplasm could lead to competition between the two amino acids for binding sites and hence to depletion of intracellular cysteine. This kind of mechanism may well account also for the mental defect, since the developing brain is particularly vulnerable to interference with the synthesis of its components.

The skeletal abnormalities and the dislocation of the lenses suggest some fault in the formation of collagen fibrils or their inter-molecular cross-linking, occasioned by the amino acid imbalance.

This case of homocystinuria is a remarkable illustration of the widespread damage to many tissues which may stem from a defect in a single enzyme in the patient's liver.

Questions

1. How can you account for the fact that homocysteine, a normal intermediate in methionine metabolism, is not present in the blood of control subjects in a loading test (Fig. 32)?

2. How could the incorporation of homocysteine into proteins be prevented?

Hyperammonaemia

Case History

Mary M., aged 9 months, had developed normally until she was weaned at 6 months. She was put on a mixed diet and promptly became irritable and less alert and she began to vomit. She was taken to hospital where she had episodes of screaming, listlessness and ataxia and failed to recognise her parents. Her condition deteriorated rapidly, her lethargy often progressing to coma, especially after a protein meal. Her head circumference was small for her height and her liver was enlarged.

Liver function tests showed slightly raised serum transaminase, indicating minor liver damage. The urine was persistently neutral or alkaline, which proved to be due to excessive NH_4^+ excretion. The presence of glutamine in the urine led to the finding of raised serum levels of that amino acid. The blood NH_4^+ was extremely high (Table 20) but fell to normal levels when the protein intake was reduced. Urea was present in the urine in reduced amount.

When Mary was put on a low-protein diet, her condition improved. The liver regressed to its normal size, the serum

Table 20. Blood ammonia and plasma glutamine

Metabolite	Patient	Normal range
Ammonia (μmol/l)	290–700	25–40
Glutamine (mmol/l)	1.9	0.55–0.70

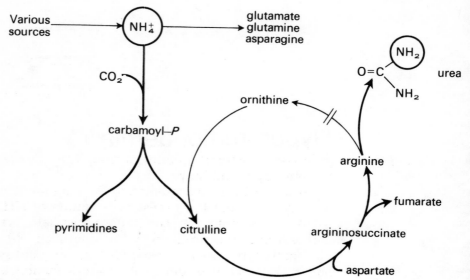

Fig. 36. Major route of NH_4^+ metabolism: the urea cycle

transaminase level returned to normal, and Mary recognized her parents once again. Because an abnormality of the urea cycle enzymes was suspected, a biopsy specimen of liver was taken at laparotomy. Shortly afterwards Mary's condition worsened, with vomiting, coma and convulsions, and excessive amounts of NH_4^+ in the blood. She died a few days later. All enzymes of the urea cycle (Fig. 36) were determined and found to be present at the normal levels, except ornithine transcarbamylase, which could not be detected.

Discussion

High concentrations of NH_4^+ are toxic to the brain and accumulation of the substance due to advanced liver disease in adults results in coma. Exactly how it exerts its toxic effect is not known, but possibly it disturbs a delicate equilibrium involving glutamate, aspartate and α-oxoglutarate (Fig. 37), on which the proper functioning of the brain may depend. (See also Chinese Restaurant Syndrome, p. 44.)

Despite the ornithine transcarbamylase deficiency, the blood urea level remained near the lower end of the normal range, presumably because circulating arginine

Fig. 37. Possible mechanisms of NH$_4^+$ toxicity

gave rise to a small amount of urea, little of which was excreted by the kidney.

An interesting consequence of the derangement of the urea cycle is that the excessive amounts of NH$_4^+$ in the tissues flood into the carbamyl phosphate synthetase pathway and thus produce pyrimidines (Fig. 36), despite the fact that they are not needed for nucleic acid synthesis and in spite of the existence of allosteric feedback control over this pathway. Excretion of orotic acid, uracil and uridine in the urine is the only way to dispose of unwanted pyrimidines.

Mary's symptoms were all referable to the toxic effect of NH$_4^+$ in the central nervous system (screaming, listlessness, ataxia, poor vision, coma) and in the liver (raised serum transaminase and hepatomegaly).

Questions

1. Why did the patient's condition not manifest itself until she was weaned?

2. What are the sources of NH$_4^+$?

Hereditary Spherocytosis

Case History

Peter T., a 23-year-old laboratory technician, was referred to hospital with suspected anaemia and jaundice. From the age of 19 he had had recurrent attacks of back pain, which was eventually diagnosed as biliary colic due to gallstones. Surgery at the age of 21, when the stones were removed, established that they consisted largely of bile pigment (bilirubin). On admission P.T. complained of malaise, his spleen was palpable and laboratory data confirmed both anaemia and mild jaundice (Table 21A).

Table 21.
Haematological data, before and after splenectomy

	A before	B after
Red blood cells ($\times 10^{12}$/l)	2.9	5.4
Reticulocytes (%)	15	1
Haemoglobin (g/dl)	8.0	15.7
Serum bilirubin (μmol/l)	78	8.5

The serum bilirubin was unconjugated and the urine contained much urobilinogen. Examination of a blood film revealed spherocytes, dark, rounded red cells, abnormally small and lacking a central pale area (Fig. 38), which suggested a possible diagnosis of hereditary spherocytosis. This was confirmed by measuring the

Fig. 38. Blood smear, showing spherocytes. For normal red cells, refer to **Fig. 42 B** (p. 70). (*Courtesy of Dr. D. I. K. Evans, Royal Manchester Children's Hospital*)

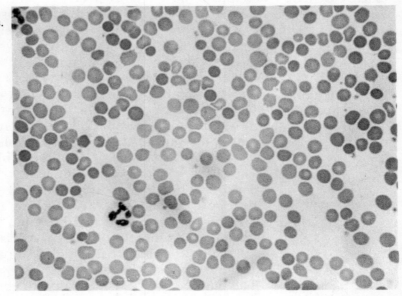

osmotic fragility of the red cells (Fig. 39), which showed that the patient's cells were more easily damaged by hypotonic NaCl solution than normal cells.

Splenectomy was performed, after which the patient felt much better and his haematological picture returned to normal (Table 21B).

Fig. 39. Osmotic fragility of red cells

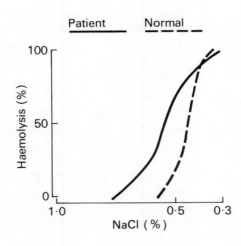

Discussion The role of the spleen is clearly of paramount importance to the clinical presentation of the disease. The organ acts as a filter, a fraction of the red cells passing through holes in the basement membrane provided that the cells are readily deformable. The biconcave disc-shaped normal red cell is sufficiently flexible to pass through, but the spherocytic cells are much more rigid and are therefore retained in the spleen, where ultimately they are lysed and removed from the circulation. This activity of the spleen not only shortens the life-span of the spherocytic red cells but the amount of haemoglobin metabolized to bilirubin exceeds the capacity of the hepatic conjugating mechanism, thus leading to accumulation of unconjugated bilirubin in the blood and the tissues, i.e. jaundice. The increased load of conjugated bilirubin processed by the liver is excreted into the bile and gives rise to urobilinogen in the intestine, some of which is re-absorbed and re-excreted into the urine and the bile.

The bone marrow is stimulated to compensate for the loss of red cells and releases immature reticulocytes into the circulation; hence the reticulocytosis. Splenectomy increases the lifespan of the red cells, although their primary defect remains. The clinical expression of the disease is thus prevented and the patient is effectively 'cured'.

The osmotic fragility test depends on the cells becoming more spherical as water passes into them from the hypotonic medium. Any additional water taken up after the spherical shape is attained results in stretching of the cell membrane and hence increased porosity, with consequent leakage of the intracellular contents. Eventually the cells burst and haemoglobin is liberated (which is measured in the medium). The test is therefore a measure of how nearly spherical the cells are before their exposure to the hypotonic saline. The spherocytic cells are smaller in diameter and they burst at a higher salt concentration.

What determines the shape of the red cell? The structural integrity and the shape depend largely on the maintenance of electrolyte gradients across the membrane, which in turn are determined by the fine architecture of the membrane and by the supply of ATP. The latter is derived solely from glycolysis, since the mature red cell lacks mitochondria and so cannot oxidize pyruvate. Na^+ ions enter the cell by passive transfer down the electrochemical gradient and are actively pumped out, with expenditure of energy, while K^+ ions are pumped into the cells against the concentration gradient, with the help of ATP, and diffuse out passively.

The spherocytic cells do not appear to lack the energy necessary for ion pumping; in fact, they consume more glucose and produce more lactate than normal red cells; the amount of glucose oxidized via the pentose phosphate pathway is similar in both (Table 22).

Table 22. Metabolism of red blood cells

Process	Patient	Normal
	μmol ml^{-1} h^{-1}	
Glucose consumed	3.5	2.1
Lactate produced	5.6	4.5
Glucose oxidised via pentose phosphate pathway	0.09	0.11

It must be assumed, therefore, that the metabolic machinery is functioning, and measurement of the rate of Na^+ efflux shows that the patient's cells are actually pumping out more Na^+ than normal cells (Fig. 40). Such increased Na^+ pumping implies a faster diffusion of Na^+ ions into the spherocytic cells, a conclusion borne out by the experiment depicted in Fig. 41, in which the amount of

Fig. 40. Sodium efflux from red cells: ^{24}Na in medium during incubation of pre-labelled cells

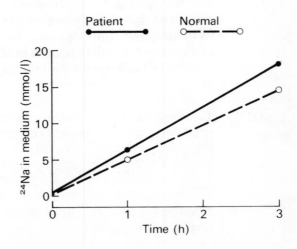

Fig. 41. Sodium gain and potassium loss during incubation at 15 °C

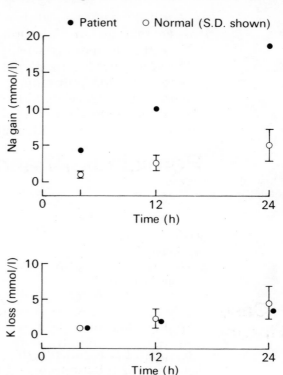

Na$^+$ gained by the cells and the K$^+$ lost from the cells during 24-h incubation in glucose medium is measured. The patient's cells gain more Na$^+$ than normal cells, i.e. their Na$^+$ permeability is unduly high, whereas the K$^+$ permeability is normal. This raised permeability to Na$^+$ necessitates an accelerated glycolytic activity to enable the Na$^+$ pump to compensate for the faster influx. The nature of the defect remains to be established, but there is evidence that it resides in a membrane protein, the faulty structure of which accounts both for the Na$^+$ leak of the red cells and their ready destruction by the spleen.

Questions

1. How do you account for the occurrence of gallstones in this patient? Are they likely to recur after his splenectomy?

2. Why do the patient's cells pump out more Na$^+$ than normal cells?

3. What is the function of the pentose phosphate pathway in the erythrocyte?

Pernicious Anaemia

**Case
History**

Herbert B., aged 55, was referred to hospital with complaints of loss of weight, weakness, shortness of breath,
sore tongue, difficulty with swallowing and epigastric
pain. On examination his skin was lemon yellow and his
tongue was shiny and smooth (Plate 10). He had pro-
gressively lost his appetite and had taken mainly liquid
foods for some weeks to avoid abdominal pain. He
complained of numbness and tingling of the hands and of
palpitation. His temperature was 39.5 °C.

The red cell count was low and the gastric secretion was
small in volume and contained no hydrochloric acid
(Table 23). A blood smear showed unusually shaped and
large red cells (Fig. 42) and the marrow contained
basophilic megaloblasts, which suggested a diagnosis of
pernicious ('megaloblastic') anaemia. The patient was
given vitamin B_{12}, which led to a complete haematological
and clinical remission.

Table 23. Laboratory
data

	Patient	Normal
Red blood cells ($\times 10^{12}$/l)	1.9	5
Gastric secretion		
Volume (l/24 h)	0.3	2.5
pH	7.0	1.5

Fig. 42. Blood cells, showing macrocytosis and poikilocytosis (A) and normal cells (B). (*Courtesy of Dr D. I. K. Evans, Royal Manchester Children's Hospital*)

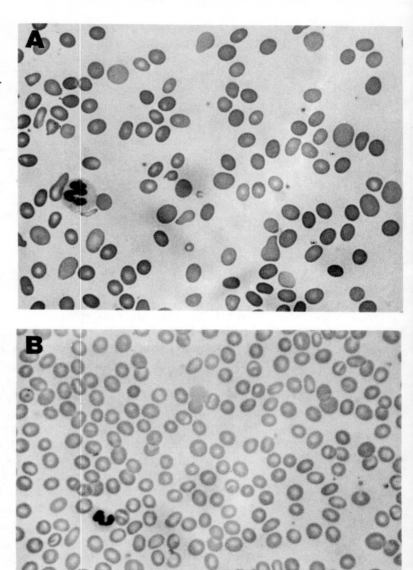

Discussion

Pernicious anaemia is due to either atrophy or surgical removal of the gastric mucosa, the secretion from which contains a glycoprotein capable of specifically binding vitamin B_{12}. The vitamin (cobalamin) is labile in the free state and is stabilized in the plant and animal world by a

Fig. 43. 'Intrinsic factor' and cobalamin absorption

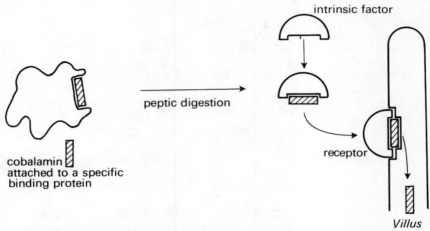

Diet

Gastric mucosa

intrinsic factor

peptic digestion

receptor

cobalamin attached to a specific binding protein

Villus

specific protein from which it is liberated by peptic digestion. The glycoprotein secreted by the gastric mucosa, formerly known as Castle's 'intrinsic factor', not only protects cobalamin from destruction in the gastro-intestinal tract but also helps with the absorption of the vitamin by the intestinal mucosa (Fig. 43). Absence of gastric secretion therefore results in failure of absorption and hence deficiency of cobalamin.

Like other B vitamins, B_{12} acts as a coenzyme. It occurs in the body in the form of either deoxyadenosyl- or methylcobalamin.

How can we account for the variety of signs and symptoms of pernicious anaemia in terms of the cobalamin deficiency? The sore tongue, difficulty with swallowing and epigastric pain, leading to anorexia and loss of weight, are clearly referable to some abnormality of the upper alimentary tract, suggesting a failure to renew the epithelium. A similar failure of erythropoiesis manifests itself in the macrocytosis and poikilocytosis evident in Fig. 42. The abnormality of these two processes suggests that the deficiency of cobalamin affects particularly the rapidly proliferating tissues. One would, therefore, look for an involvement of the coenzyme with reactions specifically geared to cell multiplication.

Only two reactions are definitely known (in man)

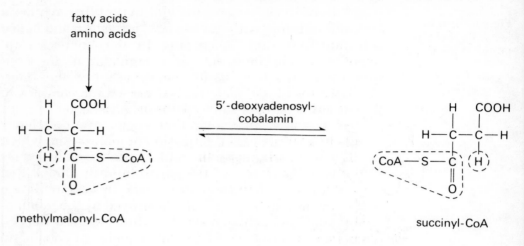

Fig. 44. Cobalamin as coenzyme for methylmalonyl mutase

Fig. 45. Cobalamin as a coenzyme for homocysteine transmethylase. FH_4 = tetrahydrofolate

to require a cobalamin coenzyme: (1) 5-deoxy-adenosyl-cobalamin participates in the conversion of methylmalonyl-CoA to succinyl-CoA, a reaction involved in the metabolism of odd-numbered fatty acids and in feeding methionine, isoleucine, threonine and valine into the Krebs cycle (Fig. 44); (2) the other reaction, requiring 5-methylcobalamin, is the transmethylation between methyltetrahydrofolate and homocysteine to produce methionine and tetrahydrofolate (Fig. 45). It is not apparent how a failure of reaction (1) or a relative deficiency of methionine due to failure of reaction (2) could have such a drastic effect on cell multiplication. The answer may lie in the other participants of reaction (2), methyltetrahydrofolate and tetrahydrofolate. Tetra-hydrofolate is readily converted into metabolites which serve as donors of one-carbon units in the biosynthesis of purines and thymine nucleotides. It has been suggested that, in the absence of the cobalamin-catalysed reaction, methyltetrahydrofolate acts as a 'trap', diverting the active

CH (NH$_2$).COOH
|
CH$_2$
| + 5'- MeFH$_4$ methyl-
CH$_2$ cobalamin $\longrightarrow$ FH$_4$ +
|
S—H

homocysteine

CH (NH$_2$).COOH
|
CH$_2$
|
CH$_2$
|
S—CH$_3$

methionine

folate derivatives from purine and pyrimidine synthesis and so interfering with nucleic acid formation and hence cell multiplication. According to the 'methyl trap' hypothesis cobalamin acts as a regulator of the concentration of active folate coenzymes, but the interrelationships of the folate derivatives are too complex to permit any testing of the hypothesis at present.

It is possible, therefore, that cobalamin deficiency results in a failure of cell formation or maturation, which would affect particularly the rapidly proliferating tissues of the bone marrow and the gastro-intestinal tract. Red cells of unusual shape and size appear in the circulation and since they are more readily destroyed by the reticulo-endothelial system than normal erythrocytes, low red cell counts and jaundice are frequent sequelae of cobalamin deficiency. Similarly, a defect in renewal of the epithelial cells in the buccal and gastric mucosa accounts for the smooth, sore tongue (loss of papillae) and for the other clinical observations pertaining to the alimentary tract.

The numbness and tingling of the hands is a consequence of degenerative processes in the spinal cord and possibly in peripheral nerves, the precise mechanism of which is still unknown.

Questions

1. Why was the patient's skin yellow?

2. What metal is part of the B_{12} molecule? Would a dietary deficiency of that metal ion result in pernicious anaemia?

Orotic Aciduria

Case History John B., aged 15 months, was referred to hospital for investigation for mental retardation and suspected anaemia. On examination he could not sit or stand unsupported, he was weak and apathetic, and his physical and mental development appeared to be retarded. He was anaemic (Table 24) and a blood smear showed poikilocytosis (Fig. 46). The most notable feature was a heavy precipitate of fine needle-shaped crystals, which developed in the patient's urine on keeping (Fig 47), about 1.5 g of the substance being excreted in 24 h. The crystals were identified by the ultraviolet absorption spectrum (Fig. 48) and the acidic and other properties as orotic acid. A small sample of liver was obtained by percutaneous biopsy and the concentrations of orotidylic pyrophosphorylase and orotidylic decarboxylase (Fig. 49) were determined: both were about 20 per cent of the normal tissue level. A diagnosis of orotic aciduria resulting from a partial enzyme deficiency was made and hence it was concluded that some of the patient's symptoms might be due to 'pyrimidine starvation'. Treatment with oral uridine,

Table 24.
Haematological data

	Patient	Normal range
Haemoglobin (g/dl)	6.9	14–18
Red cells ($\times 10^{12}$/l)	2.8	5

Fig. 46. Blood smear, showing poikilocytosis. (*Courtesy of Dr J. E. McIver, Manchester Royal Infirmary*)

Fig. 47. Crystals deposited from patient's urine. (×50)

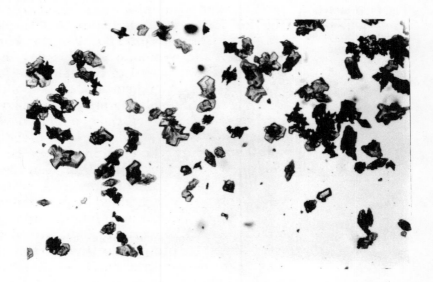

2 g/day, was instituted. John's haemoglobin level returned to normal within a few days, the urinary excretion of orotic acid dropped to less than 500 mg/day (normal 1–2 mg) and his general health improved. Six months later his height and weight were within the normal range, but his I.Q. remained at 80. His treatment was continued for several years, during which he remained well, until at 7 years of age he moved to another area.

Fig. 48. Ultraviolet absorption spectrum of urine crystals

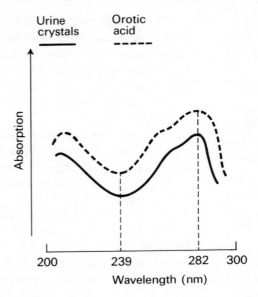

Fig. 49. Orotidylic pyrophosphorylase and decarboxylase. E_5 = orotidylic pyrophosphorylase; E_6 = decarboxylase

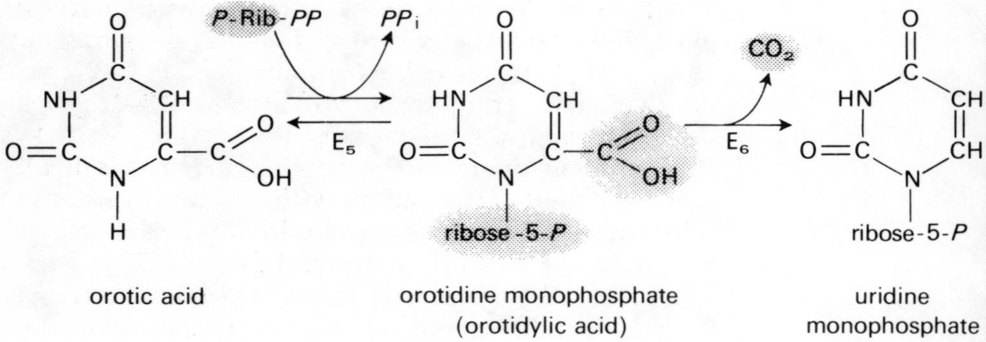

orotic acid   orotidine monophosphate (orotidylic acid)   uridine monophosphate

Discussion

The primary abnormality responsible for the patient's symptoms is clearly an enzyme defect. Unlike several other cases of inherited enzyme deficiencies described earlier, John has two enzymes, the tissue levels of which are abnormally low. In fact, orotic aciduria is probably the only human disease known so far in which two sequential enzymes are involved. How can we account for the double defect? Four possible explanations have been advanced (for a full account see reference below).

(1) A single protein, with two active sites catalysing the two reactions, and expressing a single gene, may be structurally abnormal as a result of a point mutation.

(2) A genetically determined defect of one enzyme could produce a lowering of the activity of the other enzyme by a failure of the product of the first enzyme to stabilize the second enzyme.

(3) A double mutation in two structural genes, however closely linked, is not likely but cannot be excluded on the basis of present knowledge of human genes.

(4) The regulation of gene expression, pertaining to the two genes coding for the deficient enzymes could be faulty, thus leading to reduced enzyme production.

The reactions catalysed by the two defective enzymes (Fig. 49) are the last in a sequence, starting from simple precursors, glutamine and HCO_3^-, and leading to UMP and other pyrimidines required for the synthesis of RNA and DNA as well as of various coenzymes, e.g. UTP, UDP-glucose, CDP-choline. The deficient orotidylic pyrophosphorylase and decarboxylase must, therefore, be expected to result in a serious shortage of pyrimidines ('pyrimidine starvation') and hence in a variety of biochemical disturbances whose ultimate manifestation will be a retarded physical and mental development. Nucleic acid synthesis will be impaired and the cell cycle will be deranged. The poikilocytosis as well as the anaemia testify to such disturbances in the erythropoietic system.

There is no evidence that orotic acid *per se* is deleterious and treatment of the patient with uridine, by satisfying all biochemical requirements, would thus be expected to lead to improved general health and growth. This was borne out in John B., except with respect to brain function, which was irreparably damaged at a crucial period of development.

The massive excretion of orotic acid requires an explanation. The intermediate accumulates because it can be utilized only to a limited extent, determined by the activity of the two deficient enzymes of the pathway. However, excretion dropped to one-third on administration of uridine, an observation which strongly suggests a feedback regulation of synthesis of the precursors of orotic acid. In micro-organisms two mechanisms exist for regulating the production of pyrimidines: (1) the nucleotide end-products repress the operator gene, so preventing expression of the structural gene(s) under its control, i.e.

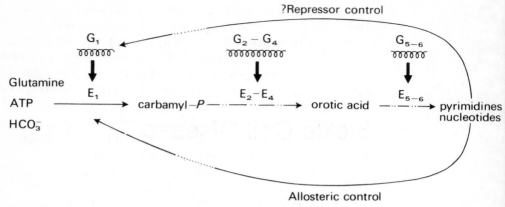

Fig. 50. Feedback control of pyrimidine synthesis. G_1–G_6 = genes of pyrimidine pathway; E_1 = carbamyl-P synthetase; E_2–E_4 = other enzymes of pathway; E_{5-6} = orotidylic pyrophosphorylase and decarboxylase

aspartate transcarbamylase, the first enzyme unique to pyrimidine biosynthesis, is not produced; (2) the pyrimidine nucleotides allosterically inhibit the enzyme already in existence. The two mechanisms represent a coarse and a fine control of the whole pyrimidine biosynthetic pathway. It is not known whether the repressor regulation is applicable to higher organisms, but allosteric inhibition of transcarbamylase or of the preceding enzyme, carbamyl-phosphate synthase, has ample experimental support.

One or both of these mechanisms presumably account for the ability of uridine to reduce the production of orotic acid by restoring a measure of feedback control, which failed in the untreated patient for lack of pyrimidine nucleotides (Fig. 50).

Further Reading

Smith, L. H., Jr, Huguley, C. M., Jr & Bain, J. A. (1972) in *The Metabolic Basis of Inherited Disease* (Stanbury, J. B., Wyngaarden, J. B. & Frederickson, D. S., eds), p. 1025, McGraw-Hill, New York.

Questions

1. What is the biochemical basis of the use of uridine in this patient? Would you expect treatment with cytidine to be effective?

2. In micro-organisms aspartate transcarbamylase is allosterically stimulated by ATP and inhibited by CTP. What is the biological advantage of this dual control?

Sickle Cell Disease

Case
History

Angela P., the second child of Jamaican parents, was brought to the Children's Hospital when aged 3. She had been excessively tired and sleepy for several months and had had frequent attacks of headache and abdominal pain. On examination the sclerae were yellow, the abdomen was distended and the spleen enlarged. The urine contained abnormal amounts of urobilinogen. The Van den Bergh test for water-soluble (conjugated) and insoluble (unconjugated) bilirubin indicated the presence in the serum of unconjugated bilirubin; the red cell count and the haemoglobin were low (Table 25). A fresh smear of blood

Table 25.
Haematological data

	Patient	Normal range
Red blood cells ($\times 10^{12}$/l)	2	5
Haemoglobin (g/dl)	4.8	14–18
Serum bilirubin (unconjugated)	+++	±

contained a few crescent-shaped red cells, (Fig. 51), but after a 24-h incubation of a sealed wet smear, all the red cells assumed the shape of sickles.

A diagnosis of sickle cell disease was made.

Fig. 51. Blood smear, showing sickle cells. (*Courtesy of Dr D. I. K. Evans, Royal Manchester Children's Hospital*)

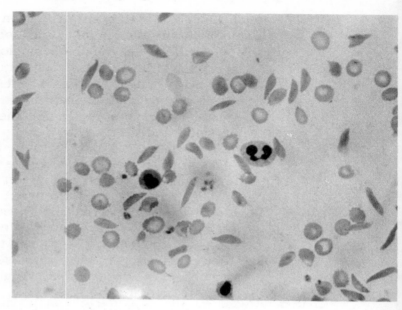

Discussion

'Sickling' occurs only at low oxygen tension and therefore became widespread only after the smear had been kept in anoxic conditions for 24 h. It could be shown that the cells assumed their normal shape on re-exposure to O_2. Pauling and his collaborators had discovered in 1949 that the red cells from patients with sickle cell disease contained a haemoglobin which differed from normal adult haemoglobin ('haemoglobin A'; HbA) by its charge and could therefore be separated from it electrophoretically. It was called 'haemoglobin S'; HbS. This discovery represents a milestone in biochemical and medical research and from it

Fig. 52. Electrophoresis of haemoglobins from family P.

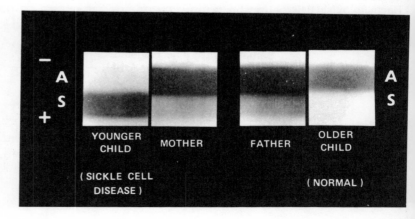

– A S +

YOUNGER CHILD   MOTHER   FATHER   OLDER CHILD

A S

(SICKLE CELL DISEASE)   (NORMAL)

much new light has been thrown on the structure of haemoglobin and other proteins and their genetic control.

Paper electrophoresis of the haemoglobin obtained from Angela, her healthy older sister and her parents (Fig. 52) showed clearly that Angela's cells contained HbS and her sister's only HbA, whereas their parents' contained both types of haemoglobin, confirming the parents' hetero-zygous status and the daughters' homozygous sickle cell and normal status respectively. In the heterozygous car-riers the normal and the abnormal gene code for their own products and both proteins are present in the red cells, a combination compatible with normal behaviour of the cells *in vivo* and hence with substantially unimpaired health. The cells do sickle, however, on extreme deoxy-genation *in vitro* ('sickling trait'). It is interesting to note that they contain 2.5–5 times more haemoglobin A than haemoglobin S, possibly owing to a greater rate of syn-thesis of the normal variety.

The clinical abnormalities of sickle cell disease can be explained by the physical properties of haemoglobin S. A concentrated solution of HbS, unlike HbA, forms on reduction a fibrous precipitate: reduced HbS is 50 times less soluble than reduced HbA. In the oxygenated state both haemoglobins are equally soluble. Sickling results, *in vivo* or *in vitro*, when the cells containing haemoglobin S are exposed to a low oxygen tension and the HbS is precipitated. Sickled cells, because of their poor deform-ability, are liable to be filtered out and removed by the spleen, which reduces their average lifespan and leads to anaemia. So much haemoglobin is liberated that the liver is unable to metabolize it. Unconjugated bilirubin accumulates in the serum and the tissues, and so accounts for the yellow colour of the sclera (jaundice). Since it is insoluble in water, it cannot be excreted in the kidney. The large quantities of bilirubin handled by the liver result in formation of more than the usual, small amounts of urobilinogen, which can be detected in the urine by special tests (Fig. 53).

Ingram submitted HbA and HbS to tryptic digestion and separated the short-chain peptides by chromatography and electrophoresis. He was able to demonstrate that the two haemoglobin hydrolysates differed only in a single peptide of 8 amino acids, one of which was glutamic acid in HbA and valine in HbS, the other 7 being identical. Since glutamic acid is a dicarboxylic acid whereas valine

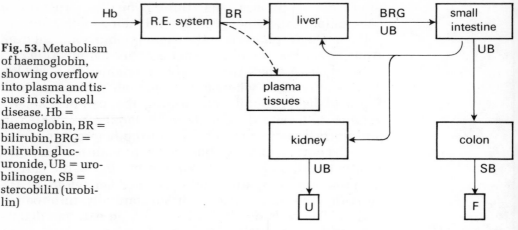

Fig. 53. Metabolism of haemoglobin, showing overflow into plasma and tissues in sickle cell disease. Hb = haemoglobin, BR = bilirubin, BRG = bilirubin glucuronide, UB = urobilinogen, SB = stercobilin (urobilin)

is neutral, the substitution increases the net negative charge of the haemoglobin, thus enabling HbA and HbS to be separated by electrophoresis. In this brilliant study Ingram showed for the first time how a gene mutation can cause substantial changes in a protein leading to widespread clinical manifestations, all resulting from the substitution of a single amino acid in a protein containing hundreds.

Angela's descent from negro stock is significant. The gene for HbS is widespread in Africa and on the Mediterranean littoral, despite the high death rate among those homozygous for the disease. One must, therefore, assume that the loss of the HbA gene is balanced by a biological advantage enjoyed by the heterozygotes, compared to the homozygous normal population with two HbA genes. It seems that the multiplication of *Plasmodium falciparum* in the red cells of infected children is inhibited in the presence of HbS: the parasite count remains low and the infection is mild and of short duration. It has been estimated that death rates from *P. falciparum* malaria among normal (HbA) children of 10 per cent and nil among heterozygotes would account for the high gene frequency of HbS—up to 40 per cent—in the malarial regions.

Questions

1. How is the bilirubin transported in the blood?

2. In what circumstances would you expect to find conjugated bilirubin in the plasma? Would it be excreted in the kidney?

Suxamethonium Sensitivity

Case History Keith P., a teacher aged 55, had been under observation by his doctor for a suspected carcinoma of the bronchus. He was admitted to hospital for a bronchoscopy. He was given thiopentone followed by 50 mg of suxamethonium intravenously as a muscle relaxant. Some uncoordinated contractions of muscle bundles and groups were followed within 30 s by the expected paralysis and respiratory arrest. Manual artificial respiration was instituted and the bronchoscopy was carried out and successfully completed in a few minutes. No abnormality was detected.

The paralysis induced by suxamethonium normally lasts 2–6 min, but after 10 min there were no signs of spontaneous respiration returning. The patient was therefore put on a respirator where he remained, totally paralysed, for 5 h. Weak diaphragmatic movements then indicated that the paralysis was beginning to wear off and after a further 90 min K.P.'s respiration was normal and he could return to the ward. He was discharged the following day, after his serum cholinesterase and dibucaine number had been determined. Both were found to be low (Table 26).

Table 26. Serum cholinesterase in patient

	Patient	Normal range
Cholinesterase (I.U./l)	8	60–120
Dibucaine number (%)	20	75–85

Fig. 54. Comparison
of the structures of
acetylcholine and
suxamethonium

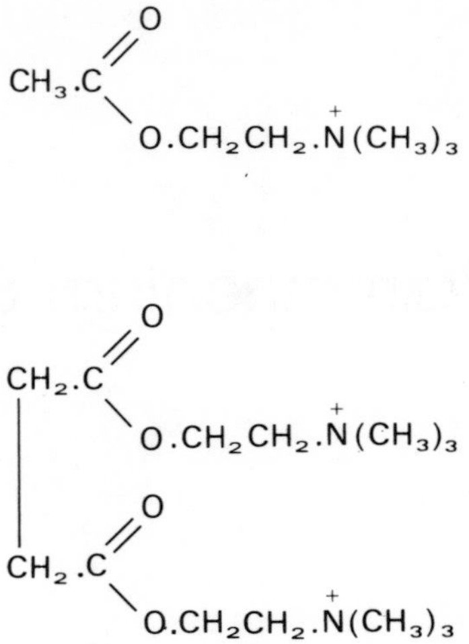

$$CH_3.C \overset{O}{\underset{O.CH_2CH_2.\overset{+}{N}(CH_3)_3}{\diagup}}$$

acetylcholine

$$CH_2.C \overset{O}{\underset{O.CH_2CH_2.\overset{+}{N}(CH_3)_3}{\diagup}}$$
$$CH_2.C \overset{O}{\underset{O.CH_2CH_2.\overset{+}{N}(CH_3)_3}{\diagup}}$$

suxamethonium
(succinylcholine)

Discussion

Suxamethonium, or succinylcholine, is a quaternary ammonium compound closely resembling acetylcholine (Fig. 54). The drug was designed to produce temporary muscle paralysis in intubation procedures. It was originally modelled on d-tubocurarine, the active principle of curare, the arrow poison used by the South American Indians. Arrow heads, dipped into an extract made from the stems of plants belonging to the genus *Chondodendron*, would carry their lethal charge into the victim's body there to cause complete paralysis and death. Tubocurarine, in carefully controlled doses, is employed by anaesthetists as a muscle relaxant in major operations, but for manipulations and minor surgery the shorter-acting suxamethonium or similar acetylcholine-like drugs are used.

The reason for the shorter activity of suxamethonium is that the drug is rapidly hydrolysed to inactive succinic acid and choline by cholinesterase present in the serum, whereas tubocurarine is not attacked. Thus it is not surprising to find that the duration of paralysis is roughly proportional to the serum level of the enzyme. The patient's serum cholinesterase was very low, which accounts for his failure to clear the administered dose of suxamethonium at a normal rate.

The enzyme is secreted by the liver into the plasma where its function is obscure, unless it be to hydrolyse any choline esters which might be generated accidentally or escape hydrolysis in various parts of the body, and which might interfere with cholinergic transmission in the muscle end-plate and elsewhere. Individuals with low levels of cholinesterase are unaware of the deficiency, which manifests itself only on administration of choline esters in quantities not arising naturally. (The acetylcholinesterase located in the motor end-plate and elsewhere has no functional connection with the serum enzyme.)

How does suxamethonium produce paralysis? Like acetylcholine itself, the drug attaches to the specific receptor protein in the muscle end-plate and in so doing opens Na^+ and K^+ channels, through which the respective ions pass, depolarizing the membrane. Whereas acetylcholine, liberated from vesicles in the motor nerve close to the receptor site, is rapidly destroyed by adjacent acetylcholinesterase, thus limiting the persistence of the agonist · receptor complex, the suxamethonium concentration at the motor end-plate is maintained by the large reservoir in the blood. The depolarizing action of the drug, and hence the paralysis, persists until the blood level has fallen sufficiently for the drug · receptor complex to dissociate.

The activity of serum cholinesterase, like that of other enzymes, is under genetic control. An abnormal gene codes for a structurally and possibly functionally faulty enzyme. It is apparent from Table 27 that the abnormal

Table 27. Comparison of normal and abnormal serum cholinesterases. [Adapted from the paper by R. O. Davies, A. W. Marton, and W. Kalow, and reproduced by permission of the National Research Council of Canada from *Can. J. Biochem. Physiol.* 38, 545–551(1960)]

Serum cholinesterase	Substrate tested	K_m	Theoretical maximum rate of hydrolysis achieved by 1 ml serum
		mmol/l	μmol ester/min
Normal	acetylcholine	1.4	4.2
	butyrylcholine	0.9	10.1
Abnormal	acetylcholine	9.0	1.3
	butyrylcholine	1.7	2.3

enzyme differs from the usual form both with respect to the affinity for two substrates and to the theoretical maximum rate of hydrolysis achieved by 1 ml serum. The atypical enzyme is also less strongly inhibited by dibucaine (a compound with a tertiary nitrogen, used as a local anaesthetic by virtue of its anti-cholinesterase

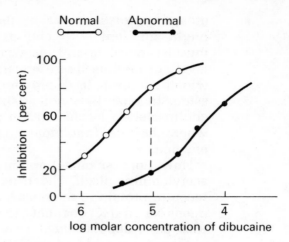

Fig. 55. Inhibition of normal and abnormal cholinesterases by dibucaine. [Reproduced by permission of the National Research Council of Canada from *Can. J. Biochem. Physiol.* 35, 339–346 (1957).] Substrate: 0.05 mM benzoylcholine. The broken line marks the difference between the normal and abnormal enzymes at 0.01 mM dibucaine, the concentration used for the determination of the 'dibucaine number'

properties) (Fig. 55). The distance between the two curves is 1.26 log units and represents an 18-fold difference in susceptibility. Thus the percentage of inhibition produced by 0.01 mM dibucaine, termed the 'dibucaine number', is an indication of the type of cholinesterase present in a specimen: normal, abnormal or a mixture of the two (Table 28).

Table 28. Normal and abnormal serum cholinesterases and their incidence

Cholinesterase	Dibucaine number	Incidence in population
Normal (N)	75–85	
N + A (heterozygotes)	50–70	1 : 30
Abnormal (A)	5–35	1 : 3600

The functional differences between the normal and the atypical cholinesterase must be due to their structures differing in some way. In fact, the enzymes can be separated by electrophoresis or by chromatography on an anion-exchange column by virtue of a greater negative charge on the normal cholinesterase. If a specimen of serum from a heterozygote is submitted to electrophoresis, the distribution of enzyme activity would reveal two distinct species (Fig. 56).

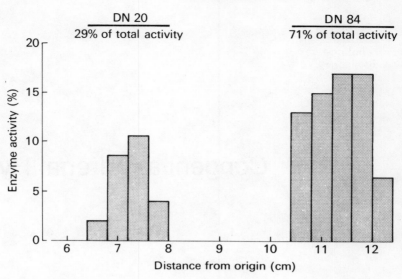

Fig. 56. Cholinesterase activity in eluates obtained from 0.4-cm strips cut serially after paper electrophoresis of serum from a presumed heterozygote (dibucaine number = 55). [From Lidell, J., Lehmann, H., Davies, D. & Sharih, A. (1962) *Lancet* i, 463]

Questions

1. What effect would liver disease have on a patient's serum cholinesterase and on the dibucaine number?

2. How can you explain (i) a dibucaine number of 50 in a heterozygote, (ii) a range of 50–70 (Table 28)?

3. Is the inhibition by dibucaine of the competitive or non-competitive type? How could this be verified?

4. How do you explain the uncoordinated contractions of the patient's muscles immediately after administration of suxamethonium?

Congenital Adrenal Hyperplasia

Case History Jane S. was born in hospital after a normal pregnancy and delivery. On examination her external genitalia were masculine in appearance: the clitoris was hypertrophied (Fig. 57) and the labia majora were bulbous and rugated, resembling the scrotum. Labial fusion concealed the vaginal introitus but was not so extensive as to cause the urethra to traverse the 'phallus'. There was a small orifice at the base of the clitoris.

Since adrenal hyperplasia was suspected to be the cause of the masculinization, a 24-h urine specimen was collected for determination of the steroid excretion pattern (Table 29A). A buccal smear for sex chromatin was positive and thus confirmed the patient's gender. Also, X-ray examination of the abdomen, after injection of a radio-opaque dye into the external genital orifice, demonstrated the presence of a normal vagina, uterus and Fallopian tubes.

Table 29. Urinary steroids: (A) before treatment, (B) during steroid therapy. ACTH could not be detected in the plasma during steroid therapy.

Steroid	Patient A	B	Normal range
	mg/24 h		
17-Oxosteroids	22.9	0.5	<1
Pregnanetriol	19	0.3	<0.4
Tetrahydrocortisol	<0.1	12.3	0.5–2.5

Jane was kept under observation. At 2 weeks she refused her feeds, vomited, became lethargic and showed signs of moderate dehydration. There were no indications of an infection and her stools were normal. She was given glucose/saline, but after 3 days she was in a state of collapse, with weak pulse and cold perspiration, signs of peripheral vascular failure. Her serum Na^+ was very low

Fig. 57. External genitalia of patient Jane S.

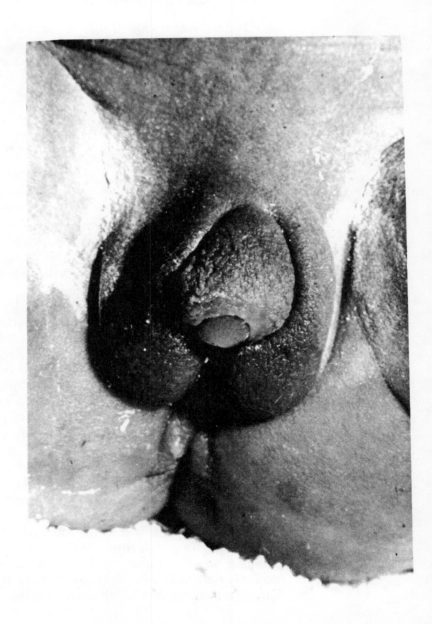

Table 30. Serum electrolytes

Electrolyte	Patient	Normal range
	mmol/l	
Na	120	136–144
K	6.4	3.5–5.0

and the K^+ was raised (Table 30) and she was in negative Na^+ balance, her urinary output exceeding the intake (Table 31). Her adrenals secreted aldosterone at a fraction of the normal rate, whereas the anterior pituitary appeared to be hyperactive (Table 32).

Table 31. Sodium balance

	mmol/24 h
Na intake	10
Na output in urine	15

On the basis of the patient's steroid excretion pattern, the deficient aldosterone secretion and the consequent Na^+ loss, and in the light of the masculinization, a diagnosis of salt-losing adrenal hyperplasia due to 21-hydroxylase deficiency was made.

Jane was treated with parenteral saline to correct her electrolytes and was given 25 mg cortisol and 0.1 mg of the mineralocorticoid 9*a*-fluorohydrocortisone. Her condition improved and after 3 days she was put on milk feeds with added salt. Oral steroid therapy was continued. A re-assessment of the adrenal and pituitary activity revealed a completely changed pattern (Table 29B).

She was discharged and thereafter continued to make good progress. Surgical repair was left until later.

Discussion

The remarkable masculinization in this neonate is clearly the result of interference with the normal development of the external genitalia. *In utero* they remain identical in the two sexes up to the 8th week, after which differentiation will begin in conformity with the chromosomal and gonadal sex. In males the formation of testicular tissue and, in particular, the interstitial (Leydig) cells play an important role from the 7th week of intra-uterine life: their secretion, testosterone, causes fusion of the urogenital slit and development of the scrotum and the penis. In contrast, the internal genital apparatus differentiates in response to a testicular peptide rather than a steroid hormone. In animals exposure of the female fetus to testosterone has been shown to lead to masculinization only of the external genitalia while the uterus and the Fallopian tubes remain unaffected.

In this patient testosterone has been secreted not in carefully proportioned doses by a developing testis but by a hyperactive adrenal cortex, stimulating masculinization where none was required. Does the normal adrenal, in both sexes, produce male sex hormone (testosterone)? Why is the cortex hyperactive in this case? How is the loss of salt related to the developmental abnormalities?

To answer these questions we must turn to the biosynthetic pathways by which the adrenal cortex produces its hormones and to the mechanism by which they are controlled. There are at least three substances whose hormonal activity is known and substantially understood; others have been proposed, but experimental evidence is still inadequate. Of the three, cortisol and aldosterone have well-recognized functions as regulators of intermediary and electrolyte metabolism respectively. The third, Δ^4-androstenedione, like its reduction product testosterone, can act as a precursor of oestrogens and thus the adrenal gland can produce both male and female sex hormones. What their physiological functions are, is a matter for speculation.

All these steroids are derived from cholesterol (which is itself synthesized from acetate) within the adrenal cortex under stimulation by adrenocorticotrophic hormone (ACTH) or angiotensin (Fig. 58). The feedback loops, which regulate carbohydrate and Na^+ metabolism, are apparent in this figure. The enzyme system stimulated by adrenocorticotrophic hormone in the cortex is probably 'cholesterol desmolase', which effects the conversion of cholesterol into pregnenolone (simplified to step A in Fig. 59). To produce cortisol and aldosterone the cortical tissue has to perform three basic operations on this parent compound: (1) to oxidize $C_3 - OH$ to $C_3 = O$, (2) to shift the double bond from the 5–6 to the 4–5 position, thus generating two conjugated double bonds, and (3) to introduce OH groups in various positions. The order in

Fig. 58. The adrenal cortex in metabolic regulation

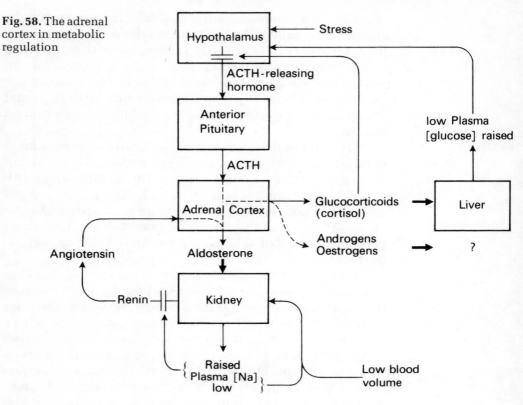

which these reactions are carried out is variable to some extent. In the slightly simplified scheme of Fig. 60 the first two of these operations are combined into step B.

Fig. 59. Formation of pregnenolone

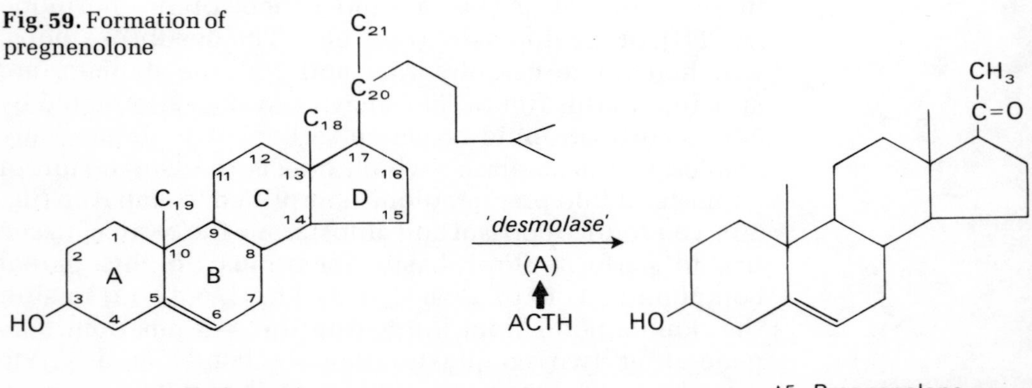

Reactions C to F are catalysed by individual hydroxylases, designated by the number of the carbon atom at which the OH is introduced. In contrast, the production of androstenedione from 17(aOH)-progesterone (reaction G) involves the removal of the side-chain by oxidation. Further oxidation at C_{19} removes the 'angular' methyl group and aromatizes ring A, an alteration which imparts oestrogenic activity to the steroid. Reduction of $C_{17} = O$ to C_{17}—OH yields testosterone (reaction H) or the corresponding 17(OH)-oestrogen.

The patient, Jane S., clearly has a genetic defect, which results in the virtual absence of 21-hyrdoxylase. Consequently, reaction D (the introduction of OH at C_{21}) cannot be carried out and little aldosterone and cortisol are produced, which is evident from the measured secretion rate of the former (Table 32) and the excreted tetrahydro metabolite of the latter (Table 29). Failure of aldosterone production leads to excessive excretion of Na^+ by the kidney, a negative Na^+ balance (Table 31) and hence low concentration of Na^+ in the plasma (Table 30). The inability to synthesize cortisol, on the other hand, results in a loss of feedback control on the hypothalamus and thus in continuous stimulation of the anterior pituitary, which secretes inordinate amounts of adrenocorticotropic hormone (Table 32). The adrenal cortex responds to the

	Patient	Normal range
Aldosterone (μg/24 h)	6	>30
ACTH	+++	+

Table 32. Aldosterone secretion rate and plasma ACTH

hormone by producing more cortisol precursors: progesterone and its 17(OH) derivative accumulate and give rise to androgenic and other 17-oxosteroids, which are excreted in large quantities (Table 29). Other metabolites of C_{21}-methyl steroids, e.g. pregnanetriol, are produced in excessive amounts (Table 29) and thereby testify to the metabolic block at the C_{21}-hyroxylation step. Suppression of the abnormal urinary steroids by cortisol therapy is further confirmation of the diagnosis: the administered hormone inhibits adrenocorticotropic hormone secretion and thus turns off steroid synthesis in the adrenal (Table 29). Excessive androgenic steroid production results not

Fig. 60. Biosynthesis of adrenal steroids

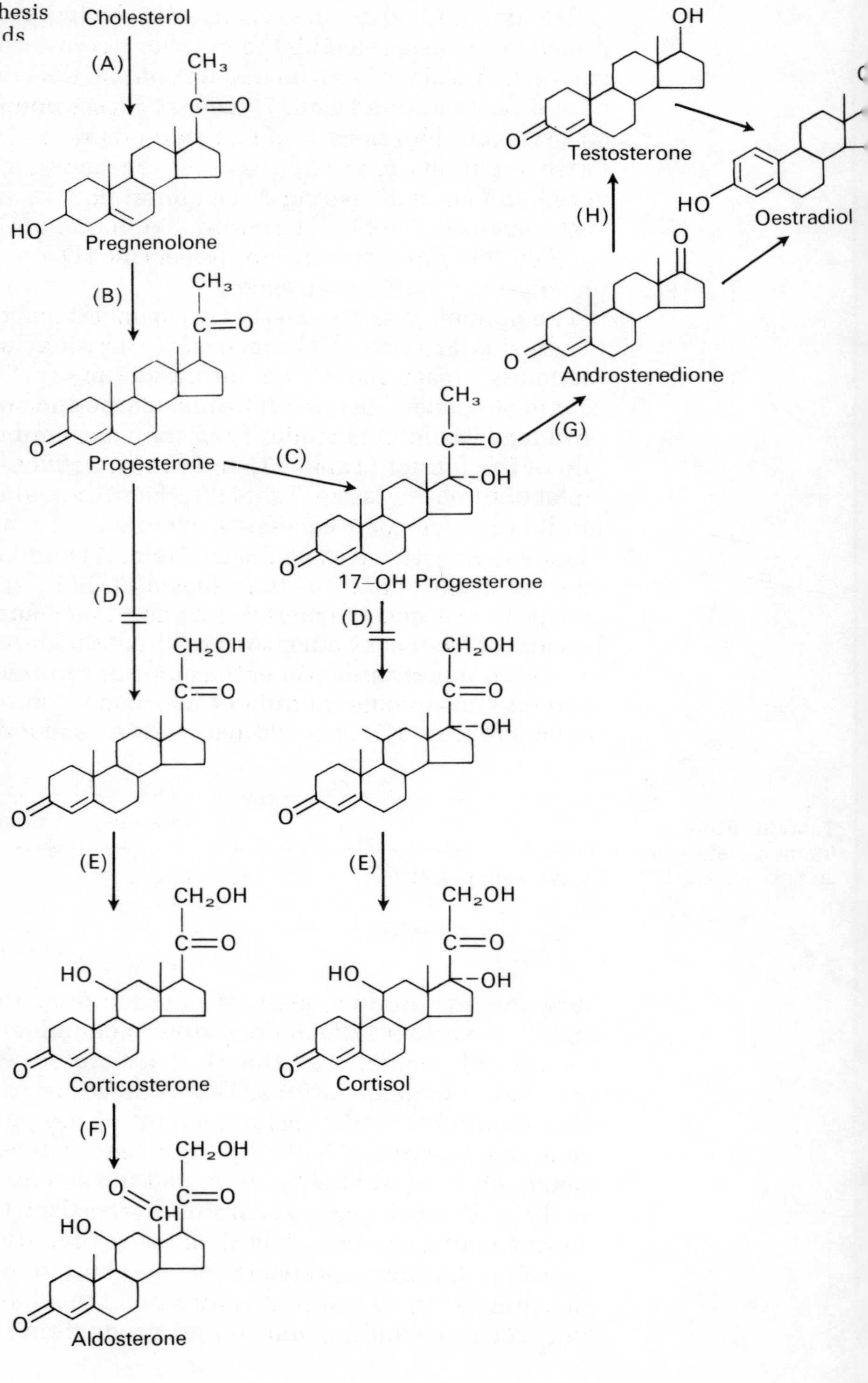

only in masculinization but also in an abnormally rapid rate of growth and epiphyseal maturation which, if not arrested early, lead ultimately to stunting.

Treatment with cortisone and an aldosterone-like drug will enable Jane to develop to full sexual maturity, with cyclic gonadotrophin secretion, normal menses and ovulation and the potential of motherhood.

Further Reading   A. M. Bongiovanni (1972) in *The Metabolic Basis of Inherited Disease* (Stanbury, J. B., Wyngaarden, J. B. & Frederickson, D. S., eds), p. 857, McGraw Hill, New York.

Question   In what would you expect the urinary steroid pattern to differ from the case described if the patient was deficient in 11-hydroxylase? (For clinical differences see reference quoted above, p. 873.)

Pseudohypoparathyroidism

Susan K., aged 12, was referred to hospital because of tetany. Since she was 4 years old she had had several generalized convulsions and she had suffered frequent attacks of muscle cramps for which she had been treated with oral or parenteral calcium salts and vitamin D.

Susan had a round face, her body build was short and stocky with a thick, short neck, and she was grossly overweight. She had a number of skeletal abnormalities, including bilateral shortened 4th metacarpals and apparently 'absent' 4th knuckles as well as exostoses in various sites.

On X-ray examination the bone density was found to be normal. There were multiple abnormalities in tooth formation, e.g. unerupted molars and hypoplasia of enamel. There was evidence of calcification in the subcutaneous tissue and in the basal ganglia.

The serum calcium level was below and the phosphate above the normal range (Table 33). The urine contained little calcium.

Table 33. Serum calcium and phosphorus levels

Element	Patient	Normal range
	mmol/l	
Ca	1.0	2.1–2.7
P	2.6	0.7–1.4

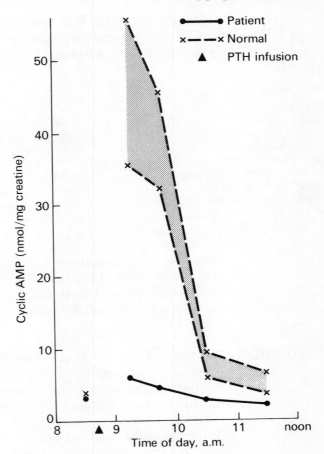

Fig. 61. Urinary excretion of cyclic AMP after infusion of parathyroid hormone (PTH). Shaded area shows normal response (mean ± S.E.M.). [After Chase, L. R., Melson, G. L. & Aurbach, G. D. (1969) *J. Clin. Invest. 48*, 1832]

In the light of this evidence a diagnosis of hypoparathyroidism was considered, but when Susan's plasma parathyroid hormone (PTH) was determined, it was found to be considerably above the normal level.

Measurement of the patient's urinary excretion of cyclic AMP in response to infused parathyroid hormone (Fig. 61) suggested a failure of the hormone to stimulate adenyl cyclase and supported the earlier observation of a more than adequate secretory activity of the glands. Their response to higher calcium levels was assessed by measuring the circulating parathyroid hormone before and after an infusion of calcium gluconate. It can be seen from Table 34 that when the serum calcium was raised to 3 mmol/l, the patient's parathyroid glands responded in a normal manner and ceased to produce the hormone.

It was concluded that Susan's primary abnormality lay not in her parathyroid glands but in the target tissues which failed to respond to the hormone produced. A

Table 34. Parathyroid hormone production at low and normal serum calcium levels. After Chase, L.R., Melson, G. L. & Aurbach, G. D. (1969) *J. Clin. Invest.* 48, 1832.

Serum component	Patient		Normal range
	before 3.5 h after infusion of Ca		
Ca (mmol/l)	1.9	3.0	2.1–2.7
PTH (ng/ml)	3.0	not detectable	<0.5

diagnosis of pseudohypoparathyroidism was made. She was given vitamin D_2, 100,000 I.U./day, which raised her serum calcium concentration to 2.7 mmol/l and lowered the phosphorus concentration to 1.45 mmol/l.

Discussion

The roles of calcium in metabolism and in other functions of the cell are manifold. The ion plays an important part in regulating the activity of certain enzymes and the permeability of cell membranes, in determining neuromuscular responsiveness, in muscular contraction and in glandular secretion and, of course, in bone formation. The calcium concentration of the blood must, therefore, be maintained within fairly narrow limits, a task accorded to three separate but interdependent physiological mechanisms operating on three levels: absorption from the gut, excretion by the kidney and dissolution of bone mineral. The mediators of this homeostatic control are three hormones: parathyroid hormone, calcitonin and 1, 25-dihydroxycholecalciferol, the latter being a derivative of vitamin D, often regarded as a hormone in view of its mode of action and other similarities with the steroid hormones.

Parathyroid hormone is secreted by the parathyroid glands, in response to hypocalcaemia, and it raises the serum Ca^{2+} concentration until it reaches about 2.7 mmol/l, when further production of the hormone ceases. By means of this feedback mechanism the concentration of the extracellular fluid calcium is maintained within the limits demanded by physiological needs. The hormone acts at all three sites: it increases reabsorption of Ca^{2+} by the renal tubule, it stimulates dissolution of bone mineral, and it aids, possibly indirectly, the absorption of the metal from the gut.

Parathyroid hormone is the messenger which carries the command to increase serum calcium (and lower inorganic phosphate) from the sensor to the effector tissues. Since it is a polypeptide (its biologically active portion consists of 34 amino acid residues), the hormone cannot enter the cells of the target tissue: a second messenger is required to transmit the message through the plasma membrane and to set in train a variety of adjustments in cellular and membrane function to achieve the desired effect. Fig. 62 indicates how the membrane-bound enzyme adenyl cyclase, functionally linked to the highly specific receptor protein, responds to parathyroid hormone by catalysing the synthesis of cyclic AMP—the second messenger.

The evidence for this role of the nucleotide in mediating the action of parathyroid hormone is threefold: (1) urinary excretion of cyclic AMP increases within a few minutes of infusing parathyroid hormone in normal subjects (Fig. 61); (2) *in vitro* the adenyl cyclase of isolated rat kidney tubules or of fetal bone tissue is stimulated within a few seconds of the addition of parathyroid hormone at concentrations as low as 5 nM (other tissues are unresponsive); (3) administration of the dibutyryl derivative of cyclic AMP (which penetrates cell membranes more easily than the natural nucleotide) to parathyroidectomized animals simulates the action of parathyroid hormone in raising the serum calcium and lowering the phosphate levels.

Fig. 62. Generation of the second messenger by adenyl cyclase. PTH = parathyroid hormone, R = PTH receptor, AC = adenyl cyclase, PM = plasma membrane

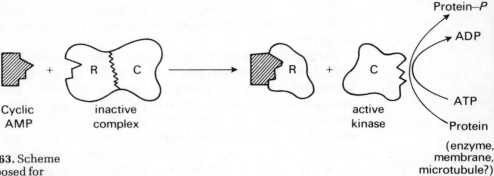

Fig. 63. Scheme proposed for intracellular action of cyclic AMP. R = regulatory subunit, C = catalytic subunit

How cyclic AMP increases calcium reabsorption in the kidney or dissolution of mineral in the bone is not known. By analogy with the action of vasopressin on the renal medulla, which is also mediated by cyclic AMP, and of other polypeptide hormones, we may speculate that the cyclic nucleotide causes phosphorylation of one or more proteins in the plasma membrane and possibly elsewhere in the cell. Translocation of calcium could be enhanced by activation, through phosphorylation, of a specific carrier and of calcium-activated ATPase.

The mechanism of action of cyclic AMP in other systems involves combination of the nucleotide with a protein complex consisting of an inactive catalytic subunit and a regulatory subunit. As the cyclic AMP binds to the regulatory subunit, the catalytic subunit is liberated and able to catalyse the phosphorylation of one or more cellular proteins (Fig. 63). The concentration of cyclic AMP in the cytosol rapidly declines as a result of the action of a hydrolytic enzyme, phosphodiesterase; the regulatory subunit · nucleotide complex dissociates, the free regulatory portion recombines with the kinase and a phosphatase removes the phosphate(s) from the cellular protein(s). In this manner the action of cyclic AMP and hence of the hormone is curtailed, so enhancing the acuity of the feedback regulation.

In the light of these mechanisms we can now interpret the clinical observations on Susan K. An inherited defect of the parathyroid hormone receptor protein and the resultant failure of adenyl cyclase in bone and renal tubular cells to be stimulated by parathyroid hormone lead to hypocalcaemia and hyperphosphataemia. The low serum calcium level accounts for the increased neuromuscular irritability (cramps, tetany, convulsions) and for hyperactivity of the parathyroid glands in response to the feedback mechanism. Thus the patient presents with the clinical features of hypoparathyroidism, yet has high levels of circulating parathyroid hormone. Soft tissue calcification and the skeletal abnormalities are more difficult to explain and may be due to hyperphosphataemia interfering in some way with normal calcification and bone development.

Further Reading

Potts, J. T., Jr & Deftos, L. J. (1974) Parathyroid hormone, in *Duncan's Diseases of Metabolism, Endocrinology,* p. 1300.

Questions

1. How do you explain the specificity of hormones, in view of the fact that the action of many of them is mediated by cyclic AMP?

2. How could you investigate whether adenyl cyclase is defective in the patient?

Thyrotoxicosis

Pamela A., a 38-year-old housewife, consulted her doctor with complaints of tiredness, nervousness and palpitation. Over several months she had become aware of a muscular weakness, manifesting itself in difficulty in going up stairs or even in brushing her hair without resting. She had lost 15 lb (7 kg) in weight, despite a greatly increased appetite and food intake, and had become irritable and jumpy, her hands showing a distinct tremor and her skin feeling hot and damp. Her pulse rate was 110/min and she complained of breathlessness and of ocular discomfort. Her eyeballs were protruding and she had a small, firm, diffuse swelling in her neck.

The general practitioner made a diagnosis of Graves' disease and referred her to hospital for confirmatory tests.

On examination the patient had tachycardia and dyspnoea. Palpation of the neck disclosed a small, diffuse goitre. Her proximal muscles were wasted. The eyes showed poor convergence and pronounced stare due to retraction of the upper eyelid (Fig. 64). Laboratory investigations yielded the information given in Table 35.

These findings confirmed the diagnosis of Graves' disease (thyrotoxicosis). An antithyroid drug (a derivative of thiouracil) was prescribed and the patient was sent home. Her condition improved slowly over the next few months and the dose of the drug was gradually lowered until clinical examination indicated that a 'euthyroid'

Table 35. Laboratory data

	Patient	Normal range
[131]I uptake in 24 h (% of dose)	69	approx. 40
Protein-bound iodine (μg/dl plasma)	12	4–8
Triiodothyronine (T$_3$) (μg/dl plasma)	1.2	0.12–0.16

Fig. 64. Appearance of patient P.A., when seen in hospital

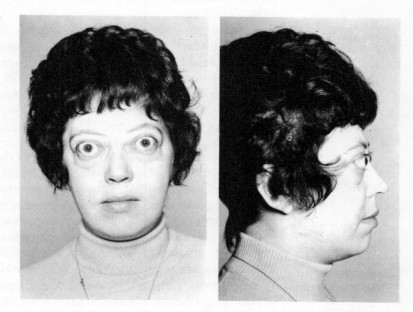

state had been reached. P.A. was maintained on this dose for another 12 months and remained well after the therapy was stopped.

Discussion

The patient had several of the clinical and biochemical signs and symptoms of a hyperactive thyroid gland. How does the hormone produced by the gland elicit the numerous effects which characterize its activity in the normal or in the hyperthyroid state? What are the pathways of biosynthesis of the hormone and how are they controlled? These questions are fundamental to an understanding of the pathogenesis of thyrotoxicosis.

The thyroid gland elaborates two hormones, tyrosine derivatives, which differ from each other only in the iodine content and are designated T$_3$ and T$_4$ (triiodo- and tetraiodothyronine, or thyroxine). The requirement for iodine is met by a highly efficient mechanism for concentrating the halogen some 40-fold in the gland. Thus iodide ingested in the diet is rapidly taken up from the plasma and converted to iodinated tyrosyl residues, both processes being regulated by the thyroid-stimulating hormone (TSH) secreted by the anterior pituitary. Production of thyroid-stimulating hormone itself is under the neuro-endocrine control of the hypothalamic 'thyroid-releasing factor'.

The iodine-concentrating power of the thyroid is the basis of the radioactive iodine uptake test: a dose of $^{131}I^-$ is given by mouth and 24 h later the radioactivity is counted over the thyroid gland. A hyperactive gland will take up iodide much more quickly and hence accumulate more than the normal 40 per cent of the administered dose, the remainder being excreted (Table 35).

Transport of iodide against a concentration gradient does, of course, require energy and is dependent on the Na^+ pump. The concentrating mechanism is within the epithelial cell itself and I^- then diffuses to the cell–colloid interface where it is oxidized to free iodine or possibly to I^+. This process is mediated by peroxidase, a haem protein which also catalyses the subsequent incorporation of iodine into tyrosyl residues of thyroglobulin (Fig. 65). (A similar system is used for labelling proteins with radioactive iodine.) Thyroglobulin is the major component of 'colloid', the fluid filling the follicles. Some 20 per cent of its tyrosyl residues are iodinated to the mono- and diiodo-derivatives, which are hormonally inactive. Two appropriately situated residues undergo coupling while

Fig. 65. Oxidation of iodide and formation of iodinated tyrosyl residues by peroxidase. TSH = thyroid-stimulating hormone

(a)

$$2 I^- + 2H^+ + H_2O_2 \rightleftharpoons I_2 + 2 H_2O$$

(b)

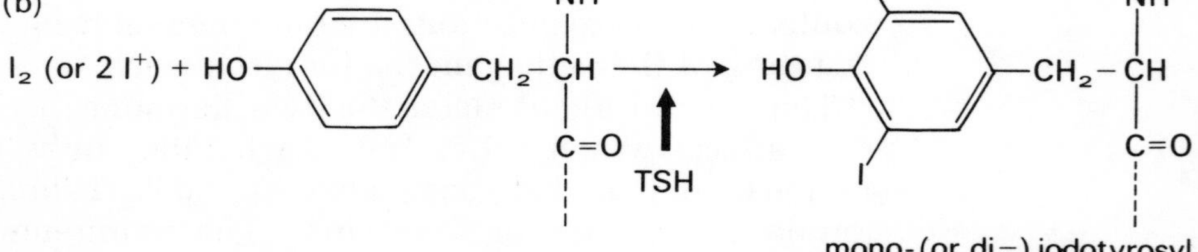

mono- (or di−) iodotyrosyl residue in thyroglobulin

Fig. 66. Coupling of iodinated tyrosyl residues within the thyroglobulin chain and formation of free thyroxine. TSH = thyroid-stimulating hormone

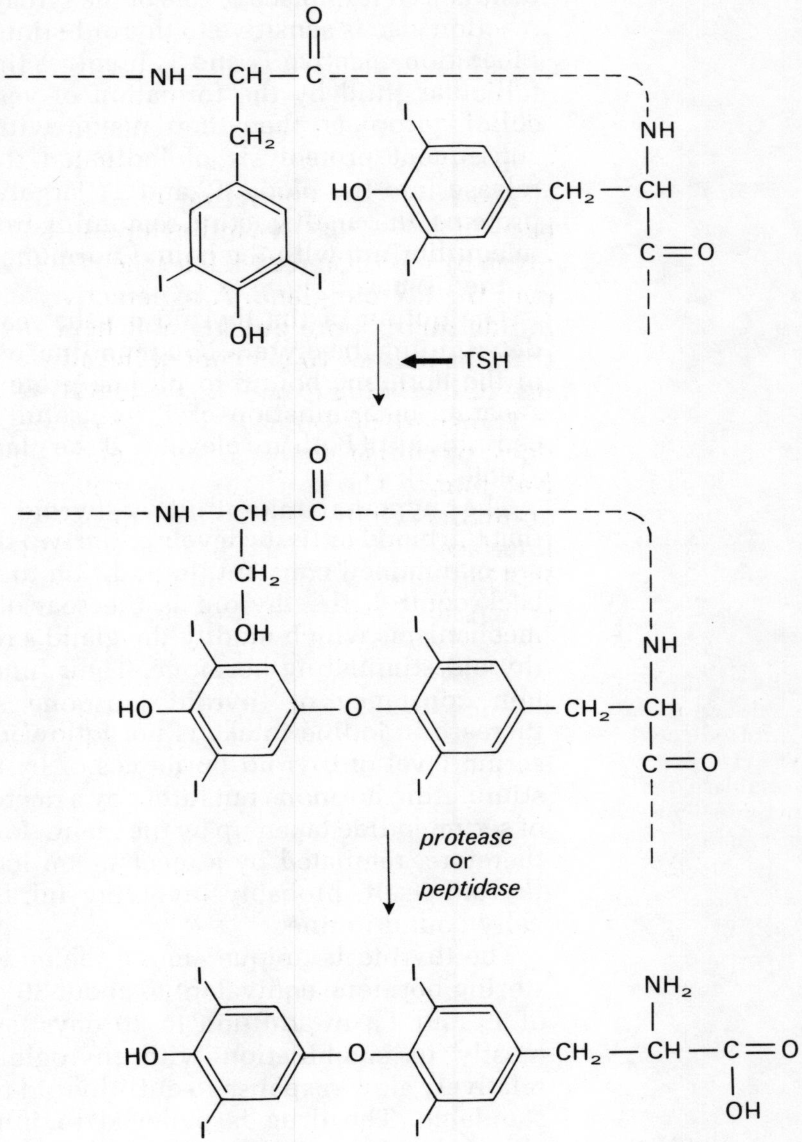

Thyroxine (T₄)

still within the thyroglobulin chain, so forming the ether link which characterizes the thyronines, and a serine residue is left in place of one of the tyrosyls (Fig. 66). This reaction also is sensitive to thyroid-stimulating hormone. Liberation of active T_3 and T_4 involves first endocytosis of follicular fluid by the formation of vesicles containing colloid droplets, then their fusion with lysosomes and subsequent proteolysis of iodinated thyroglobulin. On release into the blood T_3 and T_4 largely bind to plasma proteins, the small fraction remaining free in solution and in equilibrium with the bound hormones being available to the tissues.

The output of the thyroid can be readily assessed by determining the protein-bound iodine, which is a measure of the hormone bound to plasma proteins. Sometimes a separate determination of T_3 is useful. The serum concentrations of both are elevated if the gland is hyperactive (Table 35).

The hypothalamic-pituitary-thyroid system ensures that the blood or tissue levels of the two thyroid hormones are maintained constant. In addition to this classic feedback control, the thyroid is the seat of auto-regulatory mechanisms which modify the gland's responsiveness to thyroid-stimulating hormone. Their function is to maintain constancy of thyroid hormone stores. Thus an increase in iodine intake is not followed by a rise in the serum level of thyroid hormones or by a fall in thyroid-stimulating hormone but rather by a decreased proportion of serum iodide taken up by the gland. Iodide transport is, therefore, regulated by a mechanism located within the thyroid itself, probably involving inhibition by organically bound iodine.

The thyroid is unique among the endocrine glands in storing hormone equivalent to about 30 days' production of T_3 and T_4, in addition to 20 days' supply of iodine, mostly in combination with thyroglobulin. Thus the relatively slow response to anti-thyroid therapy is understandable. The drug is believed to inhibit peroxidase, although this has not been established definitely.

Since secretion of thyroid hormone is regulated by thyroid-stimulating hormone, the hyperthyroid state could clearly be the result of excessive stimulation. However, the circulating thyroid-stimulating hormone in Graves' disease is actually well below the normal level, which strongly suggests that the feedback control is operating. On the other hand, the presence of a different, longer acting thyroid stimulator has been reported in about 50 per cent of patients with this disorder, but the role of this immunoglobulin in the pathogenesis is obscure. The hyperactivity of the gland could be due to an intrinsic disturbance.

The breathlessness experienced by the patient is due to a reduced vital capacity, mainly from weakness of respiratory muscles, which, like the general weakness, is a consequence of a negative energy balance and hence wasting. Tachycardia and palpitation are the result, partly, of the need to dissipate excess heat giving rise to the increased circulatory demands. Partly, however, the thyroid hormones exert their effect directly on the heart muscle via a specific adenyl cyclase, which is independent of and superimposed on the adrenaline-activated system. Experiments on cat tissues *in vitro* have shown that the maximum velocity of isotonic contraction is almost double in heart muscle from hyperthyroid compared to normal animals. *In vivo*, treatment with a β-blocker (which abolishes the noradrenaline-induced activation of adenyl cyclase) does not affect stimulation by thyroxine. This and other evidence leads to the conclusion that the thyroid hormones act directly on the myocardium, in addition to the similar effect of adrenergic stimulation.

The 'nervousness' of which the patient complained, may be related to an increased secretion of adrenaline. The protrusion of the eyeballs (exophthalmos) is caused by an increased bulk of material in the orbit: lymphocytes, fat, mucopolysaccharides and water, but the mechanism of this accumulation is not understood.

Stimulation by T_3 and T_4 of metabolism, catabolism as well as anabolism, accounts for the excessive heat output of the patient and hence her hot, damp skin, and for her increased appetite and food intake, which is, however, inadequate to prevent wasting of muscles and loss of weight. The calorigenic effect of thyroid hormones has been demonstrated experimentally *in vivo* and *in vitro* by an increased oxygen consumption of the whole animal or of isolated tissues. In the latter case it can be prevented by inhibitors of nucleic acid or protein synthesis, which suggests that at least some of the excess heat is generated during synthesis of new protein. Thus, both degradation and synthesis of tissue proteins increase in the thyrotoxic patient, the former predominating and leading to a negative nitrogen balance. A similar net effect on lipid degradation is reflected in a raised level of plasma fatty acids and glycerol and a lowering of cholesterol.

The absence of a well-defined 'target tissue' of thyroid hormones (in contrast to steroid and pituitary hormones) and the diversity of their effects in many tissues raise the question of the mechanism of their action at a molecular level. No single theory has yet found general acceptance. An older hypothesis, according to which thyroxine uncouples oxidative phosphorylation in the manner of dinitrophenol, has had to be abandoned, for although relatively large concentrations of free T_4 do indeed cause the P/O ratio (high-energy P formed/O consumed) to fall, more physiological concentrations do not affect it. Another concept is supported by much experimental evidence which suggests that there are binding sites specific for T_3 in the cell nucleus and that the occupation of these sites somehow facilitates the transcription of DNA, so leading to formation of new RNA and proteins. One of these is Na^+/K^+-activated ATPase in the cell membrane: as a result, more ion pumping takes place and the intracellular Na^+/K^+ ratio falls. The increased activity of the sodium pump generates more ADP and, by stimulating oxidative phosphorylation, accounts partly for the raised oxygen consumption which is observed in tissues treated with T_3.

Questions

1. Would you expect the radioactive iodine uptake test to show a reduced ^{131}I uptake during thiouracil therapy but before the euthyroid state has been reached?

2. How would you attempt to show that iodination of tyrosine takes place almost entirely within the thyroglobulin molecule? What are the biochemical and biological advantages of this mode of synthesis?

3. Explain why the physiological effects of thyroid stimulation with hypothalamic thyroid-releasing factor do not become manifest for at least 24 h.

Glossary of Medical Terms

Anorexia: absence of appetite.

Ataxia: incoordination of muscular action.

Bilirubin: main pigment of bile; breakdown product of haemoglobin. Conjugated: linked to glucuronic acid; unconjungated: free bilirubin.

Biopsy: microscopic examination of tissue taken from the living body.

Caries: dental decay.

Cataract: opacity of the crystalline lens.

Coronary thrombosis: clot formation in the coronary artery of the heart.

Dyspnoea: shortness of breath.

Erythropoiesis: formation of red blood cells.

Euthyroid: a normal thyroid state.

Exostosis: mass of bone projecting above the normal surface of a bone.

Goitre: enlargement of the thyroid gland.

Haemopoiesis: formation of blood corpuscles.

Homozygous: having received the same gene from both parents.

Infarct: area of dead tissue due to obstruction of an artery.

Jaundice: yellow discoloration of the skin and mucous membranes due to bile pigment.

Laparotomy: incision through the abdominal wall.

Macrocytosis: presence of abnormally large red cells in the blood.

Percutaneous: through the skin.

Plaque: a patch; deposit.

Poikilocytosis: presence of irregularly shaped red cells in the blood.

Ruga: wrinkle; fold.

Sclera: outer membrane of the eyeball.

Sequela: abnormal state following a disease and due to it.

Splenectomy: surgical removal of the spleen.

Tachycardia: abnormal rapidity of the heartbeat.

Index